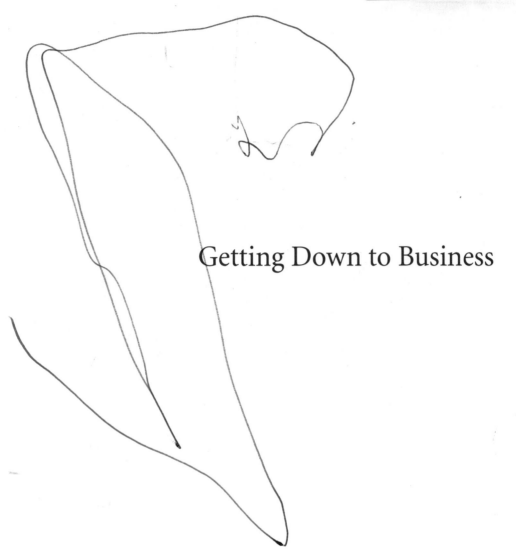

Getting Down to Business

Getting Down to Business

Successful Writing at Work

DeaAnne Kirschman

NEW YORK

Kirschman, DeaAnne.
    Getting down to business : successful writing at work /
DeaAnne Kirschman. —1st edition.
        p. cm.
    ISBN 1-57685-414-O (alk. paper)
    1. English language—Business English. 2. English
language—Rhetoric. 3. English language—Grammar. 4.
Business report writing. 5. Business writing.   I. Title.
    PE1479.B87 K53 2002
    808'.06665—dc21

                                                    2002003583

Printed in the United States of America
9 8 7 6 5 4 3 2 1
First Edition

ISBN 1-57685-414-0

For more information or to place an order, contact LearningExpress at:
900 Broadway
Suite 604
New York, NY 10003

Or visit us at:
    www.learnatest.com

About the Author

DeaAnne Kirschman, M.S.A., has worked in various industries around the country. She has been an intern for a California gubernatorial coampaign, an escrow officer for a real estate developer, a site manager for a call center, an account executive for a cellular phone company, and more. She holds a bachelor's degree in Political Science, with a minor in French. She earned her Master of Science in (Business) Administration from Central Michigan University, in August 1997. Her education and wide range of business experience give her a unique angle on both the administrative and managerial sides of business. She lives with her husband and three children in Yorktown, Virginia.

Contents

3 *Sample Letters* 103

Introduction

Albert Einstein had it right: "No idea is so complex that it can't be explained simply." Communicating in written format, with the endless rules and exceptions of the English language, can be a daunting task even for experts. *Getting Down to Business*, a how-to book on effective business writing, makes writing work for both novices and veterans. It also responds to the growing demand of companies for better communication in today's diverse, competitive, and fast–paced market. The need is so great that studies are being conducted and companies are taking action.

In seventy interviews with top executives from six different industries, 63% were angry every time they received e-mails that were either forwarded chain mail, or just plain baffling. A COO of a successful new E-business says that all he really wants in a written document is, "plain, simple English!" A vice president who has been in business for over forty years says that clarity and understandability play a vital role in business success: "Your written word is a direct representation of **YOU**. Make it clear. Make it good. Make sure everything with your name on it is accurate."

A vice president in the private banking industry describes some of the business correspondence she reads as, "totally lacking punch—too many unnecessary words." New York's Commissioner of Education ordered 250 top officials to take a writing class after reading too many "confusing" letters. Federal agencies, like the Securities and Exchange Commission, are implementing "plain-language" movements, ordering companies to write more clearly. The Veterans Benefits Administration has saved $500,000 a year by training its employees to write better. In short, companies are recognizing the financial benefits of good writing, and they are begging for ways to get it.

Getting Down to Business also addresses educational needs. States including California, Indiana, Nevada, and others are recognizing the need for increased student proficiency in English by mandating exit exams for high school seniors. It is estimated that up to 70% of students in some Southern California classrooms speak English as a second language. At one school in Sacramento, California, seventeen different languages are spoken. These students will become part of a budding unique workforce and will need to know how to write well in order to do well. An executive at a leading computer firm says, "I find myself continually correcting my assistant—an English major from a leading university—who is unable to write effectively." Even English majors and MBAs need refreshers, and always have room for improvement.

Getting Down to Business is designed to:
- save companies time, money, and energy
- offer readers a wide array of business writing samples
- explain grammar rules in simple terms
- help starter businesses make a professional first impression
- enable small, medium, and large corporations to communicate clearly
- boost writing confidence and skills of students
- be the first comprehensive and cohesive business writing guide of its kind
- provide practical tips for successful and positive writing style
- allow today's diverse workforce to sharpen their skills

▶ WHO NEEDS THIS BOOK?

Getting Down to Business will help students about to enter the workforce, established professionals, government agencies, educational departments, corporations of all types and sizes, colleges and universities, new businesses, and more. Each of these audiences needs to be a partner in the practice of professional business writing. This book will help you achieve a mutually successful professional relationship through effective communication. It is every organization's gift to themselves and to their employees. It is also an invaluable tool for any individual who wants to succeed in business and needs to know how to write more productively.

If you are a new college graduate, read *Getting Down to Business* from cover to cover. It will outline all of the letter-writing business basics that you will need to get started down a successful career path. It will also provide you with real-world examples of winning letters, and will offer standardized formats for you to follow to ensure effective communication. You will find helpful tips in each section that will serve as reminders about what needs the most focus.

It is imperative to understand the "rules of the road" in business writing—and you will stand out very quickly if you can express yourself well. A commander in the United States Navy offers this golden rule: "Remember the 'Fourth Law of Thermodynamics': Heat applied to you is heat *not* applied to me." This book will take the "heat" off you by arming you with all of the tools necessary to do your best writing, and make a powerful impression!

Getting Down to Business is also a precious resource if you are an established professional. As many people advance in their careers, they find that they don't have extra time to look up rules in large volumes. And they certainly don't have the time to pore over letters that drone on endlessly about nothing in particular. They need clear, concise writing. This book provides solutions for all of those issues: it takes you back to the basics with your thinking. It is a concise, compact wealth of information that allows busy professionals a quick refresher and reference guide.

Companies will benefit from using *Getting Down to Business* in the workplace. It will give their employees quick access to uniform writing skills that will help them communicate effectively, both internally and with their clients. It will also help reduce the number of mistakes, which take time to correct, and end up costing the company money. We know that time equals money in business. So, rather than spend valuable time enveloped in online research, employees can pick up this quick reference guide and readily find answers to many common writing questions.

▶ ABOUT THIS BOOK

Writing shouldn't be painful. It should be like riding a bike, where, once you get it, you can pedal with the best of them. But, just like riding a bike, you have to start somewhere. You begin with training wheels, then you move to a nice banana seat two-wheeler … then, with lots of practice, and with an understanding of the rules of the road, you find yourself cruising at top speed on your 18-speed carbon fiber racing bike. You have made it to the Tour de France! Wherever you are in your professional career, this book is dedicated to helping you succeed.

Getting Down to Business will not only help take the pain out of writing, but will teach you how to build the scaffolding necessary to create a powerful and effective business letter—a letter that will get the results you are looking for. The whole goal of this book is to provide a straightforward, comfortable, and logical framework for you to use in business writing.

The book is divided into four easy parts: Basic Training, Composition, Sample Letters, and Grammar. Basic Training is a section on fundamental writing and communication skills—the basics. This part is crucial because it is where most people get tangled. Think of it as stretching before exercise, or planning "the game" before you play it. You wouldn't dream of just running wildly out onto the field or court without a game plan, or without preparation. It is the same with writing—you need a clear plan first. In this section, you will learn how to organize your thoughts, write clearly and concisely, analyze your audience, and still manage to be yourself in your writing.

Once you have mastered these first critical elements, you can begin learning some standard "plays." Think of the Composition section as your "playbook." This section will outline the customary parts of a business letter or memo. It will also cover electronic correspondence, a topic that has become worthy of a book in itself. You will learn about international correspondence, and also how to properly begin and end a letter. You will find sample resumes and cover letters, and everything you need to know, but never thought to ask, about envelopes and paper.

In Section 3, you will find a wide variety of real-world sample letters that you can draw upon for all of your business correspondence. The sample letters range in topic, from acknowledgments to transmittals, and they provide a useful overall picture of what your finished product should look like. If you don't find the exact type of letter you are looking for, then find one with similar subject matter and adjust your letter to address your specific needs. If you apply all that you have learned from each section when writing your own letter, you should be writing successfully in no time!

Of course, you need to know the "rules of the game." Enter Part 4: Grammar. In this section of the book, you will find simple grammar rules, punctuation rules, and some commonly misspelled words. You will also learn how to cite sources properly and how to avoid plagiarism. After you have completed this section, you will be armed with solid background information on writing basics, and you will be ready to write!

Getting Down to Business is meant not only to make writing easier for you, but also to help you achieve success through quality writing. It is written to bolster your excitement about writing. Most of us recognize that grammar rules and regulations about sentence structure are not usually topics that evoke thunderous enthusiasm. So, by the time you have completed this book, hopefully you will be enlightened about the beauty and art that is involved in creating your own written masterpiece! Your writing, even in business, is the essence of *you*. Remember to think of your writing as Michelangelo thought of his statues:

"I saw the angel in the marble and I carved until I set him free."

So, start with a positive outlook, and keep in mind two fundamental things as you read this book:

- Writing is something you can learn to do well—and may even enjoy.
- Once you learn, you will gain supreme confidence in your ability, and you will succeed!

Getting Down to Business

Basic Training

▶ ORGANIZE YOUR THOUGHTS

> Writing is easy. All you do is stare at a blank sheet of paper
> until drops of blood form on your forehead.
>
> —GENE FOWLER

We have all been there: at that first moment before you begin to write. You sit down at your desk with a fresh beverage, take out a clean sheet of paper, grab a pen, sigh a huge sigh, whisper to yourself, "okay, here goes" and then . . . nothing. Blank. Nada. It happens to the best of us. And it is the hardest part about writing. But, here's the good news: Once you have gotten past those first few agonizing moments, and you begin to put your thoughts in motion, the hardest part is over! You realize that you are a person with a purpose, and you are ready to embark on your writing journey.

Organizing your thoughts before writing is absolutely critical. It is probably the single most important step in the entire writing process. Before you even sit down at your desk or computer, you have to start thinking. So, do whatever it takes to put yourself in a mental state of free thought flow . . . go for a walk, a run, or swing on a swing. Sit on the couch and eat chocolate if that invigorates your mind. If you are at the office, close your door. If you have a cubicle, like so many people do these days, then take your break alone, or have lunch with just yourself and your thoughts. The point is: you need to allow yourself the ability to really focus.

THINKING STYLES

This might sound more like a lesson in Zen Buddhism, but clear thinking makes all the difference in your writing performance. You can start by first figuring out what type of thinker you are. This seems funny, but isn't it obvious in everyday life how differently people think? Just try getting three small children and their grandmother to agree on what to have for dinner, and you will see what I mean. You could conceivably have ten people in one meeting, with each person looking at the same issue in a diametrically different way. So, you have to understand what kind of thinker you are.

There are two basic thinking styles that can be associated with writing: *linear thinking* and *free association.*

LINEAR THINKER

You are a **linear thinker** if you organize your ideas in chronological or sequential order. If you are working with a timeline, you simply list events chronologically, starting with the first event:

> **Example:** *You are having a problem with construction for your office building. The builders are not following the timeline that your company and theirs initially agreed on. In order to get your point across, you need to make a chronological list of the whole project—bulleted or numerical, from beginning to present.*

Sample Notes: **Linear thinker** (using chronology):
- Contract was signed July 4, 2001—project to be completed by July 4, 2002.
- Phase One of construction to be completed by Oct. 4, 2001. Actual completion date: Dec. 12, 2001.
- Phase Two of construction to be completed by Jan. 4, 2002. Actual completion date: March 4, 2002.
- Today is May 4—Phase Three not yet completed.
- What's going to happen with the Fourth, and Final Phase?
- List steps that need to be taken to achieve timely completion.

If you are thinking *sequentially,* you make an outline or a list that begins with your most important ideas first. You then move down your list of thoughts in descending order of importance:

Example: *You need to write a letter to all employees about a new dress code. So, you sketch a quick outline that covers what you need to say in order of importance.*

Sample Notes: **Linear thinker** (using sequence):

1. Announce new Dress Code, effective January 4, 20XX
2. List specific changes
    a. No halter tops
    b. No open-toed shoes
    c. No jeans
    d. Casual dress on Fridays
3. Describe consequences if new code not followed
4. Thank employees for cooperating
5. Give contact information for any questions

FREE ASSOCIATION THINKER

You are a **free association thinker** if you use no particular sequence in your initial thinking. You have a thought, jot it down as it comes to you, and then provide supporting details last. You might write down key words that you know will trigger your memory later. You will eventually do an outline, but you need to see all of your ideas laid out on paper first.

A pharmaceutical sales executive refers to this type of thinking as "bubble thinking." Her thoughts come to her at light speed, so she writes down her notes as quickly as she thinks of them. She then circles each separate idea in its own "bubble" so she can logically categorize them later. When she's done taking notes, she rearranges each bubble until her letter flows sensibly.

Sample notes for New Dress Code (using **Free Association**):

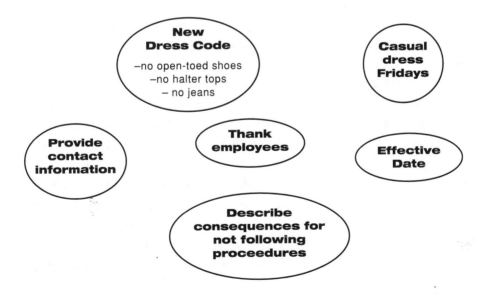

The thinking style notes in this section are obviously very brief, but they address the important points. Of course, the length of your outline will vary depending on the amount of content you need to discuss; and subcategories can be added to reinforce certain points that require specific information. The important thing to determine is what kind of thinker you are. Once you have done that, you can apply yourself to your next step: organizing your notes logically!

■

WORDS OF WISDOM
My words fly up, my thoughts remain below.
Words without thoughts never to heaven go.

—WILLIAM SHAKESPEARE

■

ORGANIZATIONAL METHODS

The most important thing to remember when organizing your thoughts is to stay focused on the big picture—be sure you are "sailing in the right ocean." Think *everything* through first. Don't bog yourself down in the details, or you will sail right off course into the abyss. There are some common organizational methods that pertain to almost every business writing scenario. Think of these standard organizational methods as your "true North." All you have to do is survey the situation, apply the appropriate method, and sail. Some common formats/techniques include:

Alphabetical method: To organize global office locations or an office supplies list, you would list them this way.

Chronological method: Company history information, meeting minutes, and corporate calendars usually appear this way.

Deductive method: Used by attorneys—often referred to as "*I.R.A.C.*": Issue, Rule, Application/Analysis, and Conclusion. You begin with a general issue. Then you state the rule of law, then how that law applies to your issue, and finally, what conclusions can be drawn—or, why your side should win. This method is also used for sales or promotional letters and is also excellent for business proposals.

Inductive method: You begin with details and examples and build up to the general issue by citing supporting evidence along the way. You might use this method if you are a teacher and you have a problem student.

Inverted Pyramid: Used for general business correspondence (letters, memos, reports), and also for writing a book. You begin with the overall layout, and then emphasize key elements in order of decreasing significance. Your goal is to bring different parties to mutually pinpointed focus—to be "on the same page."

List: This method is a basic list of the fundamentals—use either bullets or numbers. Some examples are a list of company rules, a list of needs, or anything basic.

Order of Location: This method is used to define territories or regions—often used in marketing. A sales executive for a cellular company might write a report on cellular sales volume by region. Television advertisement marketing strategies are often drawn up using order of location.

Priority Sequence: Issues are listed in order of priority, starting with the most important and ending with the least important. This method is perfect for proposing a set of steps or procedures; a construction project would be drawn up this way.

Problem/Solution: This method is used to simply state a problem and then describe how it was solved. It usually ends with a synopsis of the final outcome.

WORDS OF WISDOM
Time is your most precious resource—
take the time necessary to create an outline,
so that you write a quality letter the first time.
A commander in the U.S. Navy advises his
subordinates to divide the writing process into
thirds: one-third on brainstorming and outlining,
one-third on writing, and one-third on revising.

TIME TO OUTLINE!

Once you have decided which organizational method suits your purposes, you are ready to create an outline. A full-scale outline is only necessary when your document is longer than a few paragraphs, but it is still helpful to draw up a "plan" before you start writing.

An outline serves as an overview of your intended subject matter, and can be written in several different forms: executive summary, abstract, or the standard way, using Roman numerals, capital letters, Arabic numerals, and lowercase letters. In most cases, the standard way is the easiest way. It is the way we were taught in school, and it delineates headings and subheadings well.

Following is a standard outline written by a Senior Account Representative, describing his understanding of the facts after meeting with a client:

I.   Overview
  A. List meeting attendees
  B. Company direction
II. Discussion of Topics
  A. Electronic EOR Download Process
    1. Manual bill payment
    2. Info-systems "go-ahead" to move into production
    3. Commend all parties involved in project
  B. Claim record processing
    1. Automatic weekly update of *OUCH* system
    2. NADR system features
      a. Alleviates repetitive data entry
      b. Provides more detailed claimant information
  C. Provider File Upload Process
    1. Electronically upload provider demographic info.
    2. Requires a PPO Pricing database
    3. Complex process to be tabled pending further discussion
  D. Employer Level Bill Processing and Reporting
    1. Allows clients to receive reports at employer level
    2. This system not currently needed—table the issue
III. Use of the AMN in surrounding states
  A. Provide document outlining savings to DCHO
  B. Provide DCHO with provider directories for other states
IV. OM*NET* Leasing
  A. DCHO will lease access to OM*NET* worker's comp network
  B. Conduct further discussions regarding potential agreement
V.   Business Objectives

Your outline helps you set the direction of your letter. It forces you to weed out unnecessary words and stay focused on your goal. In an outline, you start with clear thinking, then define your major points, and finally rearrange them until they make sense. It also helps to list all supporting details or facts—subcategories—so that you have the substantiation you need in your final drafted letter.

When you have a long document, such as a proposal or a report filled with details, or a document that is packed with technical language, use an executive summary. The executive summary itself should not be long—one paragraph is usually enough. Sometimes, one really solid sentence will suffice. If you need to use a list format to offer preliminary details in your summary, you can simply state the nature of your document and then use a bulleted list to describe its contents. The point is to use the executive summary as your introduction to the lengthy issues to follow:

> *I am writing to relate our understanding of the processing issues we covered during our discussions about the* WCIS-OUCH *integration project.*

That is really all there is to it. Just that one short sentence describes twelve pages of ensuing technical information. Also remember that you don't have to state that it is the executive summary—it just *is.*

SUMMARY

We know how important it is to think before we speak—the same goes for writing. You must think before you write. Time is your most precious resource, so take the time necessary to create a quality letter by organizing your thoughts first. Clear your head, and get in the right frame of mind—this is a critical part of the organization process. This is also where most people get hung up.

So, understand what kind of a thinker you are, linear or free association. Then, begin brainstorming—let your ideas flow, jot them down, and then rearrange and revise as needed. Apply whatever organizational method best fits your needs as you move through this preliminary process. When you are ready, you can gradually channel your thoughts into an outline. If it helps, you can follow the advice of a successful law student who uses the *IRAC* organizational method: "I do a better job persuasively if I set up the 'skeleton' of my document first, and then write a limited factual synopsis. Finally, I go back and write what the rest of the law is and how the facts relate, and then revise what I originally wrote."

If you want to write a successful letter, preparation is essential. Remember to begin with a clear mind, determine your goal, and then stay focused on the core issues throughout the process. When you feel that you are on the right track with your thought process, you are ready to ask yourself . . . to whom am I writing?

▶ ANALYZE YOUR AUDIENCE

> Clients 'buy' from people they feel they can trust—in the end, it is our client relationships that win us the deals.
>
> –KEVIN KELLY, COO AT EQUARIUS, A TECHNOLOGY CONSULTING FIRM

There is no way around it: the only way to write successfully in business is to understand your audience. It is a simple theory that is put into practice in the one million subtleties of our everyday lives. If you are in a restaurant where the server takes great care to get your individual order right, makes all of the changes you requested, and even makes you feel like he is tending to you alone—you won't notice the fifteen other tables he's waiting on—you will probably give him a great tip. You both come away happy. Or, if a doctor really *listens* to you, carefully collects all of your information, and treats you with respect, you will not only remain loyal to that doctor, but you will refer every friend you have to him or her. These are both examples of successful business relationships that exist because someone took the time to understand his or her "audience."

As a writer, you want to get your point across first and foremost, so make it easy for the reader. If you are able to put yourself in your reader's shoes, you will have a mutually successful relationship, and your correspondence will be well received. Your reader will feel comfortable building a relationship with you, and you will have earned that reader's trust, loyalty, and most definitely his or her business. This applies to clients, colleagues, supervisors, and subordinates—*everyone*. Always keep in mind the common business principle that "telling is not selling." Listen, learn, and succeed.

AUDIENCE ANALYSIS QUESTIONS

Audience analysis deserves serious attention, regardless of the size and scope of your audience. And it entails more than just learning a few bland statistics, such as your reader's company and position. But you can make it easy on yourself by simply creating a bulleted list of possible questions:

- What is the reader's age, sex, present job, educational level, and past experience?
- What is the reader's primary spoken language?
- How does the reader prefer to be addressed?
- What form of business communication does the reader use most? An executive from a leading computer company says she does not look at anything but electronic correspondence—e-mailing her would be your best bet.
- What type of clientele does the reader serve?
- What is the reader's demeanor—conservative or moderate?
- At what level of authority is the reader? Can he or she act on your letter?
- What matters most to the reader in a written document? Does he prefer brevity to details? Does he have disdain for unwarranted attachments? Or, does he prefer to have everything possible sent to him?
- Does the reader have a sense of humor?
- What type of business language is the reader accustomed to? Technical lingo, medical lingo, legalese, and so on?

These are just examples of questions that will help you understand your audience. As you build your professional relationships, you will be able to dig into even finer details about your reader that will personalize and enhance your communication.

COMMUNICATION STYLE AND TONE

It is critical to use a communication style and tone that fits your audience. You wouldn't write the same thing to your best friend from college as you would to a prospective client. But, you do need to remember the "human element" in your business writing. In today's technical age, it is more important than ever to personalize and warm up your messages. So, once you have analyzed your reader, try to strike the balance between professional and "too cozy." And whatever you do, err on the side of warmth—don't be stuffy and cold. That will only impress *you*. Use simple, direct communication that is geared straight to the reader. No one will ever complain that your letters are too easy to read, or that they understood them too well.

An executive for a high-tech company says this about getting in touch with your audience:

> As a company, we make a conscious effort to cut back on the technical jargon when writing or speaking to our clients. Using jargon only makes your clients feel inferior and self-conscious. People want to be around those who make them feel good. Therefore, if we want to build lasting client relationships, we need to understand our audience and communicate at their level.

An executive vice president of a leading medical malpractice insurance company offers an excellent illustration of how important it is to understand your audience.

> We have a beautiful office building in the wine country of Northern California. We also had a serious peacock problem on the grounds outside of our beautiful building. What began as two lonely peacocks turned into a flock of cousins, uncles, aunts, and young peacock offspring.
>
> The employees took to the budding peacock family, and began feeding them and treating them as pets. Before long, we had peacocks flying to our outdoor lunch tables, even disrupting some lunch meetings. Have you ever seen a peacock fly? We knew we had to do something.
>
> So I wrote a company-wide letter, asking all employees to please refrain from feeding the peacocks, as they were not our pets. I realized halfway through my letter that, in a company of over 300 employees, there were some who were sensitive to animal rights issues. So, I had to gear my letter to the most sensitive person. If I was too blunt or careless in my tone, it could have affected employee morale, and caused bigger issues than flying peacocks. In the end, a letter that I thought would be simple to write—easy issue, not a million-dollar deal—turned out to be a good lesson in understanding my whole audience.

ANALYZING AUDIENCE CHARACTERISTICS

It is just as important to understand the thinking style of your reader as it is to understand your own. Many different theories have been studied over the years about individual personalities and how people think. And, in an age where understanding the people behind the machines is becoming increasingly important, we need to pay special attention to this subject. Carl Jung theorized that there are four basic styles of communication:

1. *Sensor/Action Style*: This person is action-oriented, very hands-on. They are driven, determined, tough, competitive, confident, and assertive. They can also be domineering, arrogant, and impersonal. Typical careers for this person include: doctor, athlete, executive, pilot.
2. *Thinker/Process Style*: This person is an information-processor. They organize and strategize, gather information. They are analytical, logical, critical, methodical, organized, persistent. They can also be insensitive and judgmental or inflexible. Typical careers for this person include: lawyer, engineer, scientist, financier.
3. *Feeler/People Style*: This person is a people-person. They are socially geared, communicative, team-oriented, warm, friendly, persuasive. They can also be subjective, overly sensitive, and overly cautious. Typical careers for this person include: teacher, psychologist, sales associate.
4. *Intuitors/Idea Style*: This person is creative, theory-oriented, driven by ideas. They are reflective, serene, "dreamers," adventurous, flexible. They can also be undisciplined with time, unrealistic, and manipulative. Typical careers for this person include: artist, professor, researcher, writer.

It is important to note that these are simply theories that Carl Jung used to try to categorize certain personality traits that he observed. You or your reader could be a combination of any of these descriptions—or, you could seemingly not fit any particular category. So, pay close attention to your *reader*, and be careful not to make personality assumptions based solely on occupation.

STYLE AND TONE WITH AUDIENCE IN MIND

Once you understand what type of communication style your reader uses, then you can decide what kind of tone and correspondence is appropriate to the situation:

Informal note:

Hey, Chris, isn't that great about Ben passing the bar? We've offered him a job as part of our litigation team—we know he'll do great. He wrote some awesome briefs as an intern.

Company newsletter:

Please join us in welcoming our newest lawyer, Ben Ammerman, to our team. Ben just passed the Bar exam with flying colors—Congratulations, Ben! We know from the work he has already done for us that he will be a valued new asset.

Press release:

The law firm of Fetzer & Kirschman announced today the hiring of Ben Ammerman to the position of associate attorney. Ammerman was a successful intern with the firm for 4 years through law school before passing the Bar Exam.

SUMMARY

With the pace of business today, you no longer have the time for "fluff"—you need to write so that people can understand you. You need to get into the mind of your reader and really tap into what will make a difference and what will solicit a positive response. This requires listening, learning, and thinking about what you want to say, and to whom you are saying it, *before* you write. If you put yourself in your reader's shoes, you will both enjoy strong communication and a better business relationship.

So, you need to "get personal." Learn what makes each person tick. Who are they? Where are they coming from? What is most important to them? When you have gathered what you need to know about your reader, use a tone and a communication style that is audience appropriate. Then decide what type of correspondence fits the situation—personal note, formal letter, press release. Remember that listening breeds trust. Once you are clear about to *whom* you are writing, then it is time to be clear about *what* you are writing.

▶ BE CLEAR AND CONCISE

. . . let every word tell.

–WILLIAM STRUNK, JR.

William Strunk said it vividly . . . " . . . let every word tell." His classic book, *The Elements of Style*, is only eighty-five pages long, and it remains one of the best books ever done on the subject of writing. He understood that he would lose his audience after the first page if he didn't write clearly and concisely. As a professor, he repeated his mantra to classrooms packed with riveted students: "Omit needless words!" He apparently said it with such force and determination that many of them never *did* forget. And neither should you.

Some of the best-written works of all time have been clear and concise. Take our Constitution's Bill of Rights, for example. It is only one page long . . . and, although the print is very small, it said enough in that single page to run one of the most powerful countries on Earth for over 200 years. Not bad. If our forefathers can lay the foundation for an entire country on one large sheet of paper, then we can certainly be clear and concise with our business correspondence and letters.

TECHNIQUES FOR CLARITY AND CONCISENESS

Is there one technique that you can use in all of your business correspondence? Yes! *Be consistent* with your clarity and simplicity. Pretend you are the captain of a sinking ship, and you only have a few sentences to get your message out to your entire crew. This means that you have to write in a way that is understandable to everyone, from the ship's cook to the navigator. An executive in the insurance industry said this about his business communication:

> "In over forty years of business, I always used the same technique: plain, simple, understandable, and to the point. You can't go wrong that way—it leaves no room for confusion or misinterpretation."

The COO for a technical company took a business writing class where the instructor had a mathematical formula for grading the writing. It was simple—those who wrote the shortest sentences and used the simplest words got the highest scores. The class learned that clarifying and simplifying their prose led to drastically increased understanding of the message content. They got the point of the message. No one got bogged down trying to decipher difficult words, or having to wade through murky sentences. So, try scoring yourself—write a sample letter and have a friend or associate read it. Ask them to score each sentence by giving ten points for every concise statement and subtracting ten points for every confusing statement. If your first score is on the low side, never fear! That only means you have room for growth and a new opportunity to make yourself an even better writer than you thought you could be.

In order to write clearly and concisely, you have to ask yourself some essential questions with each sentence that you write. In his book, *Politics and the English Language*, George Orwell suggests that you ask yourself the following key questions:

- What am I trying to say?
- What words will express it?
- What image or idiom will make it clearer?
- Is this image fresh enough to have an effect?
- [Also], could I put it more shortly?
- Have I said anything that is avoidably ugly?

BE CLEAR

There was an orator named Demosthenes, who lived in Athens, Greece in 300 B.C. He was highly praised by everyone for being a brilliant communicator because he used lofty words that made him sound intelligent. His style was animated and captivating. The people got so caught up in listening to him that they never knew what he was actually saying. One day he proclaimed another orator to be better than he. He said simply that, when the other man spoke, he spoke to the level of the crowd, and they heard his *message*. They left his speeches *knowing* what he had said. Thousands of years later, the same theory applies—be clear above all. Your most important objective is to get your message heard.

An executive vice president with over 17,000 employees under his direction, says this about clarity:

> Write like you would talk to a friend. This may be [risky], but it is clearer and establishes your style. For example, 'Charley, it is time we sat down face to face and talked this deal over," instead of, "At your earliest convenience, would you please extend your permission to arrange a mutually agreeable time to convene a business meeting to discuss . . . blah, blah, blah.

The funny thing is that "blah, blah, blah" is probably exactly what the reader would be thinking if they received the latter note. So, think like an executive vice president—a leader— and get to the point with your writing.

If you think and write like a leader, you will eventually *become* a leader. Leaders have a clear direction. They have a clear goal. And they cut a path straight to that goal. This means that it is absolutely critical to keep your "eye on the ball" throughout your writing process. If you have total clarity about what you want—and don't be afraid to get specific—then make the decision to go after it, keep focused, and sharpen your letter until it glistens like a blade in sunlight.

GET TO THE POINT

■

WORDS OF WISDOM
If you want to be clear and authoritative with
your prose, never begin a statement with
the words, "I think." When trying to persuade
someone to do something in business,
you need to *know*, and you need to be
able to communicate what you know.

■

The best thing you can do if you want to get your message across is get to the point. Say what you want to say, support it with facts, be specific, ask for what you need, thank the reader, and then end the letter. You can organize different types of letters in different ways, but following a logical order and getting to the point are imperative to being clear and understood, no matter what type of letter you are writing.

In an inquiry letter, for example, the order of the letter should be as follows: what you want, who you are and why you are asking for it, and then end it with a brief thank you. Remember that the reader is most interested in the gist of your letter. And you will come across as a straight shooter, someone who is interested in what matters most, if you write in that order.

Sample Inquiry Letter

129 Harris Drive
Seattle, WA 99223

March 2, 20XX

Mrs. McGhee
1590 Newton Road
Seattle, WA 99223

Dear Ms. McGhee,

Are you taking any interviews yet regarding your new position as Marketing Director with Metro Pharmaceuticals? If so, will you please consider giving your first interview to me? I am an established columnist with the *Seattle Times*, and I am dedicated to giving a fair and honest interview that will show you off to the world. Your time and your thoughts would be greatly appreciated.

Sincerely,

Joanne Matheson

Joanne Matheson
Staff Writer

Here are some sentences that leave the reader guessing, followed by examples of how to be specific:

Vague: I hope to meet with you again soon!
Clear/Specific: We are looking forward to an answer about that contract by January 4.

Vague: We greatly appreciated your help with the Randolph matter. You are a great new asset to our team.
Clear/Specific: Your decision to renegotiate the Randolph contract earned us an additional $10,000. Pam and Ronan tell us you are the best new attorney on staff.

Vague: Your investment should increase significantly by next year.
Clear/Specific: Your investment should increase 20% by next year.

Vague: The new system has been very profitable.
Clear/Specific: The new system has reduced operating costs by 30%.

Vague: somewhat behind schedule.
Clear/Specific: one week late.

On the other hand, here are a few powerful adverbs and adjectives that can add punch to your point:

directly involved
unflagging dedication
promptly accepted
productive meeting
hefty raise
influential employee
invaluable asset
priceless decision

DELIVER BAD NEWS CONFIDENTLY

Bad news is bad news. To deliver it confidently is to write with the idea that either a situation will be solved or that you will work through it. It may be news from a financial planner to her client that the market has gone belly-up, or it may come in the form of a supervisor needing to counsel an employee. Many executives warn not to hide behind form letters or e-mail; rather, face-to-face communication is best. However it is delivered, bad news is not made better by trying to gloss over it or manipulate words to try to confound the reader.

Of course it remains important to understand your audience in order to determine what tone is appropriate. But, in most cases, it is best to just get it out there—deal directly. You will avoid bigger trouble in the long run if you deliver a clear and honest message. And you can still be tactful, or even delicate, if the situation permits. You can even use an opening "cushion sentence" or two to ease elegantly into the message. But, in the end, it is better to give the bad news straight, and grant your reader the personal respect of being able to handle the truth.

Following is a letter written to a concerned client by an executive financial planner:

605 Main Street
New York, New York 10002

January 15, 20XX

Mrs. Fetzer
5021 Eastern Avenue
Tucson, Arizona 85708

Dear Ms. Fetzer,

This past quarter has been one of the most difficult periods for all investors. Our company and clients, unfortunately, were no exception to this broad downturn in the market. In this period of decline, our funds have continued to outpace the leading market indicators. While our investments, on the whole, declined 6.3% over the past quarter, the S&P 500 declined 10.6% and the Dow 12.1% for the same period. These are unprecedented times and they call for unprecedented investor discipline.

While mid-October marked the definitive decline of the stock market, there were signs leading up to this event that telegraphed the imminent downturn. Lack of investor confidence and chaotic economic events in the Japanese markets aggravated an already fragile American economy.

We feel that this is an excellent opportunity for investors to move into the bond market. While equity investments have taken a beating, cash flow has maintained a positive measure making quality corporate bonds an attractive investment alternative.

Now more than ever, portfolio diversification is a must for every investor. In good times and bad, diversification has been a hallmark of our company's investment philosophy. Please don't hesitate to call if you have any questions about your portfolio or the investment strategies discussed here in this letter.

Sincerely,

Joyce Wyman
Joyce Wyman

Several points are noteworthy in the above letter:

1. The writer cut to the chase. She explained some very complex issues as clearly as possible.

2. People take their money very seriously; so, she used an appropriate tone of both professionalism and authority. She essentially said, "I'm carefully studying this issue because I know it means a lot to you."
3. She was tactful and understanding, using opening language that put her on the same level as the reader—she seemed to say, "Wow, hasn't this whole thing been an unbelievable experience for all of us?"
4. She did not try to back away from the obvious gravity of the situation.
5. She was specific with her facts, using percentages and statistical data to demonstrate her research of the case.

PROVIDE SUPPORTING DETAILS

Be sure to separate fact from opinion—this means providing supporting details. Companies will make decisions based on facts and numbers, not on how you feel. So, while it is important to make recommendations, be sure that they are fact-based, and that you provide plenty of supporting evidence. Be sure to also offer *solutions* to issues, not just a long list of problems and facts. One Chief Operating Officer explains:

> I can't stand it when someone writes up a business case loaded with facts without suggesting "what to do." I think people tend to *not* offer an opinion because they're afraid of failing. I like people who are willing to take responsibility and accountability for their opinions.

DON'T HEDGE

Hedge words and phrases are used when the writer is looking for a buffer to soften his statement. It is usually done because he is either not prepared to submit something as fact, or he thinks he can dodge the issue by hiding it in extra words. Here are some common hedge words and phrases to avoid:

> according to our records
> as far as I can tell
> as per your request
> as you might know
> could
> for your information
> I wish to thank
> if I recall
> in due time
> in my humble opinion
> in the near future
> in view of

it is my understanding that
just about
likely
might
mostly
permit me to say
probably
pursuant to
with reference to

Again, two of the most important aspects of a well-written letter are clarity and candor. Take out every single word that you don't need in order to make your message clear. Be specific with facts, and use opinions sparingly, unless it is part of your job to make recommendations to clients. When you do offer suggestions, be sure to back them up with facts. Remember to say what you mean, even if it is not the best news.

WHEN YOU HAVE TO SAY NO

The biggest problem with having to say no in writing is that the readers can't see you. They can't hear your tone of voice, and they can't see your body language. They also have no chance to respond or comment before you are through with your message. You can compensate for these drawbacks by personalizing your letter:

1. *Be clear.*
2. *Be careful with your tone*—Use a gracious and thoughtful tone.
3. *Anticipate the reader's questions*—try to answer them in advance.
4. *Put your letter to the test:* Ask yourself if you would say the same thing you are writing to the reader's face? Does it sound real? In other words, don't say something like, "Please permit me to thank you for interviewing for the position, but . . . ." A better choice of words would be: "We enjoyed meeting with you to discuss our new sales position. We have decided on another candidate for this job, but we will keep your resume on file. We thank you for your time and wish you the best of luck in your job search."
5. *Put yourself in the reader's mind: How would you feel if you got this letter?* Remember that your ultimate goal is to put your best foot forward in every letter you write.

There are several standard rules to follow when you have to say no:

- *Never say no in anger:* When you write something down, it is permanent. Don't let your words come back to bite you. Control your temper and remain professional—find the words to express your dissatisfaction in a professional manner.

- *Never belittle anyone:* Don't be accusatory. If you are a credit collector, for example, and you need to collect money from someone, you should assume *nothing.* Use a respectful and understanding tone—your company will look better, and you may even get the money you are asking for.
- *Never say no carelessly:* Write to show that you care about the reader, even if you are part of a large corporation. Form letters should not be too impersonal. Remember the importance of showing people you come in contact with that you care about them individually. The last thing you want to do is make them feel like "a number."

BE CONCISE

■

WORDS OF WISDOM
"Be concise—this requires thinking and planning.
Limit the length of a business letter to
one to one-and-a-half pages, tops. Length of
a letter is important. Keep it short."

—JIM WHITTLESEY, EXECUTIVE VICE PRESIDENT

■

An executive in the insurance industry says that his favorite thing to tell colleagues and clients about writing concisely is: "I'm sorry [the letter] is so long—if I had had more time, I would have sent you something shorter." This proves that he understands the importance of organization, thinking, and planning prior to writing and sending a letter. If you haven't thought it out, you are bound to ramble. If you ramble, you lose your audience.

The executive goes on to describe the best letter he ever wrote to an underwriter after a lengthy conference call enumerating all of the reasons to stay with the client. He was essentially begging the busy underwriter to stay on board, but he needed to be concise:

> I sent a two-page letter to him. The first page was completely blank—white as alabaster, nothing on it at all. The second page had the words: "Above are all of the reasons why you should stay with this client. Regards, Bruce."

It worked. The busy underwriter was both amused and thankful for the short letter—he decided to stay with the client. The executive who wrote the letter had his reader in mind when he wrote, and the reader felt understood and appreciated. The writer accomplished his goal with a blank sheet of paper and the fewest words possible.

CHISEL AWAY NEEDLESS WORDS

Here is a list of wordy phrases, along with some options:

Rambling/Wordy	Concise
A great deal of	Much
Are of the belief that	Think that
As per your request	At your request
As you may already know	As you may know
At a later date	Later
At all times	Always
At this time	Now
Based on	Because
Despite the fact that	Although
Do an analysis of	Analyze
Equally as	Equally
Essentially unaware	Does not know
General idea	Idea
Group consensus	Consensus
Here locally	Locally
I think that we should	We should
I'd like to thank you	Thank you
In order to	To
In the area of	Approximately/about
In the course of	During
In the event that	If
In view of	Since
Inasmuch as	Since
In-depth study	Study
It is clear that	Clearly
Make a recommendation	Recommend
Month of December	December
Need something along the lines of	Need a
Over with	Over
Plan of attack	Plan
Schedule a meeting	Meet
Subsequent to	After
Take action	Act
The majority of the time	Usually/frequently
Until the time when	Until

We ask that you return the contract	Please return the contract
We can be in a position to	We can
With regard to	Regarding
With the exception of	Except

Remember that wordiness literally costs money. It costs more in paper and postage, if you are using regular mail. But, its biggest cost is time and efficiency, which, of course, is worth much more. So, even though it may take you longer to plan and edit your letter before writing it, it will benefit you in the long run. You will earn a reputation as a clear thinker, someone who will get down to business.

CHECK YOUR PARAGRAPH AND SENTENCE LENGTH

It is just as important to watch "rambling" sentences and paragraphs:

- **Paragraph:** Give the reader a break by keeping your paragraphs within about five to seven sentences. The old standard rule has always been that a *minimum* of three sentences is necessary to constitute a paragraph. But, new standards have unofficially adjusted the rule to just making sure it is not too long.
- **Sentence:** Be sure your sentences are in logical order as you build them into a strong paragraph. Watch the placement of your thoughts, and put your emphasis where you think you need it most—usually at either the very beginning, or the very end, of your paragraph. Last sentences are usually used as either summation points (of what was just written), or as tie-ins to the next paragraph. This keeps your words flowing and your reader engaged.

AVOID REDUNDANCY

While it is sometimes important to repeat ideas to get your message understood, be careful not to overdo it. Use repetition as a tool only when it helps to emphasize your point:

We care about quality. We care about lives. We care about you.

But, there is a difference between effective repetition and redundancy. Following is a list of redundant phrases and their more crisp alternatives:

Redundant	**Concise**
As a general rule	As a rule *or* generally
Begin to take effect	Take effect
Close proximity	Close
Collectively assemble	Assemble
Continue on	Continue

Contractual agreement	Contract *or* agreement
Cooperate together	Cooperate
Current status	Status
Depreciate in value	Depreciate
Endorse on the back	Endorse
Final completion	Completion
Final outcome	Outcome
First and foremost	First *or* foremost
First priority	Priority
Foreign imports	Imports
Honor and a privilege	Honor *or* privilege
Invisible to the eye	Invisible
Lose out	Lose
May possibly	May
Meet at 3:00 o'clock P.M.	Meet at 3 P.M.
Normal practice	Normal *or* practice
Other alternative	Alternative
Past history	Past *or* history
Personal Opinion	Opinion
Quick and speedy	Quick *or* speedy
Rarely ever	Rarely
Reason is because	Reason is
Refer back	Refer
Repeat again	Repeat
Revert back	Revert
This particular instance	This instance
True facts	Facts
Tuition fees	Tuition
Virtually in effect	In effect
Vitally important	Important
Whether or not	Whether

SUMMARY

The best way to ensure that your writing is clear and concise is to use Orwell's "questions to ask yourself" as a guideline. Then, think like a leader, and get to the point. Don't write anything that requires decoding, or waste time cluttering up your messages with unnecessary words. It only leads to confusion, which of course, is not your goal. What *is* impressive is someone who is able to get to the heart of a matter and affect people. Pinpoint your goals and write them down clearly and concisely. Then edit yourself and be sure you have the exact information you need in each letter—no more and no less.

Also remember that, while it is critical to be clear and concise, you need to use your common sense above all. Don't omit critical information just because you think your letter seems too long—it may be necessary to *add* information in order to clarify something. And when you have to say no, do so with tact and grace. Apply these two characteristics to everything you write, and you can't go wrong. Finally, keep in mind William Strunk's notion: " . . . let every word tell." If you make every word count, then you will save time and money, and you will be an effective communicator.

▶ KEEP IT SIMPLE

> I never write 'metropolis' for seven cents because I can get the same price for 'city.' I never write policeman because I can get the same money for 'cop.'
>
> —MARK TWAIN

Mark Twain was simple on prose and brilliant on thought. He knew that complicated words only complicate a message. And he realized that complicated messages serve no useful purpose, except to baffle the reader and leave them feeling inferior and uninterested. So, unless your intent is to stump your reader, keep it simple!

One of the great examples of simple writing is Abraham Lincoln's Gettysburg Address. Lincoln was never accused of being a long-winded speaker—if anything, he was maligned for being too simple a man. Little did his critics understand how important it is to keep things simple and focused. This is not to say that you speak or write *down* to people, but rather, that the minute you try to go over their heads, or get too complex, you have lost them. As President of the United States, Abraham Lincoln knew that his speeches needed to address everyone from "Joe, the barber," to "Joe, the judge." His theory worked.

The Gettysburg Address is only three paragraphs long, and it took only two minutes to say aloud. In those two minutes, Abraham Lincoln motivated an entire country and changed the direction of the world. Today may be the age of complex technology and global business deals/interaction, but that doesn't mean that you should complicate your communication. On the contrary—in order to communicate effectively, you need to get back to the basics. You must simplify your writing.

THE TIME FACTOR

It is vital to consider the time factor of the reader: How busy are they? Will a long note or letter frustrate them? Will they have time to respond to everything you have written? Ask yourself these questions and write accordingly—better yet, find this out about the reader ahead of time. That will help you gauge how much you should include in your letter, and it will show the reader that you consider his or her time valuable. Another way to look at it is this: the better you write, the less time your boss has to spend editing and reviewing your work!

One great way to save time in your letters is to remove the warm-up paragraph. Sometimes we like to give people background when we are telling a good story because we think it adds flavor to the tale and brings the listener into our world. That may be true if you are sitting with a friend sipping a latte on a lazy Saturday afternoon. But, if you are sending a busy executive a business message, then that is the last thing you should do (unless it is specifically requested). Instead, engage your reader in the first paragraph by providing important and relevant information in as compelling a way as possible.

A sales executive for a leading pharmaceutical company says that her clients are busy doctors who have very little time for "fluff." She understands that she needs to respect their time constraints, so she gets her messages out in a few simple sentences:

1225 James Street
Secaucus, NJ 67097

July 14, 20XX

Dr. B. Bratt
1074 North Aztec Boulevard
Secaucus, NJ 07094

Dr. Bratt,

Here is a copy of the invitation to the program for Monday, September 24th. How does the timing sound? We can have everyone sitting down and eating before Dr. Schaefer starts speaking. This way everyone can get out at a decent time.

Once I get the thumbs up from you and Dr. Schaefer, I'll print up the invitations and send them to you. Please let me know what you think.

Thanks again,

Pam
Pam

SIMPLIFY YOUR WORDS

You know that old saying about children getting to the point with their words? They don't waste any time trying to impress anyone because they already have a clear picture of their goal. If they want a cookie, for example, they simply ask, "May I have a cookie please, Mom?" (the nice version). There is really no reason the same rules shouldn't apply to doing business in adulthood. If you need a coworker to attend a meeting, for example, the best way to approach him is to simply ask, "Can you sit in on the meeting this afternoon, Bob?" Bob should not only be refreshed by your candid manner, but he should also be able to give you a direct response.

We can work with simplicity and clarity. It is the cloudiness in life that poses the biggest challenge. The point here is that, when you simplify your writing, you get to the point much faster and more effectively. So, you don't want to go overboard trying to impress someone with your cosmic vocabulary because: 1) They'll probably think you are trying to hide some portentous shortcoming; or, 2) They won't understand a word you have written, and therefore won't act on your letter.

Here are some examples of flashy, overdone words that in most situations only serve to complicate and confuse:

Flashy	Simplified
Advise	Say/tell
As per	According to
Aggregate	Total/collective
Ascertain	Determine
Cogitate	Think about
Cognizant	Aware
Comprised	Made up of
Commence	Begin
Conjecture	Think/believe

Disbursement	Payment
Endeavor	Attempt
Initial	First
Interface with	Get together with
Forward	Send
Maximal	Fullest
Modus Operandi	Method
Nominal	Small
Obviate	Make necessary
Per Diem	A day
Permit	Let
Predicated	Based
Proclivity	Tendency
Recalcitrant	Resistant to authority
Remunerate	Pay
Said	*Don't use as an adjective
Same	*Don't use as a noun
Scrutinize	Inspect
Sine qua non	Vital/essential
Subsequent	After
Transpired	Happened
Undersigned	I/me

Here is a list of outdated, stale expressions that you can replace with a more conversational, simple tone:

Stale	Fresh
A large segment of	Many
Acknowledge receipt of	Received
Allow us to express appreciation	Thank You
At this writing	Now
At your convenience	By April 8
Check in the amount of	Check for $500.00
Due to the fact that	Because
During the course of our investigation	Our investigation showed
Enclosed, you will find	Here is . . .
Give due consideration	Consider
Held a meeting to discuss	Met and discussed
I have before me the . . .	I received the . . .
In accordance with	According to

In the event that	In case/if
Made the announcement that	Announced
Please be advised that	*Just state the issue
Too numerous to mention	Numerous
Until such a time as	When
Upon completion, mail form	Mail completed form
We are engaged in the process of	We are
We regret to inform you	We are sorry

This doesn't mean you have to "dumb yourself down" to a third grade level in order to get your point across. In fact, it is best to assume the recipient is at least as intelligent as you are—so using a respectful tone is key. It also doesn't mean that you should never use any "flashy" words; if you need a flashy word to illustrate your point, then go for it. And you can certainly use more complex words if they fit the situation. But, remember that not even the smartest grammarian can remain intrigued for long by a windy, pompous diatribe, however well placed the "big words" are.

ELIMINATE "BUZZWORDS"

"Buzzwords" are trendy business terms that take up unnecessary space in your letters. Newer professionals tend to use them because they think it somehow validates them as professionals, and that their bosses will take them more seriously if they use them. But, in truth, using buzzwords is rather like wearing bell-bottoms—at some point, the trend will shift, and you will be left out in the cold with no original thoughts of your own. So, start now by developing your own writing technique—*sans* the buzzwords—and pave your way toward simpler, easier writing:

Buzzwords	Normal Words
Concinnity	Harmony
Functionality	Effectiveness/success
Guesstimate	Estimate
Incent	Inspire
Meaningful	Actual, real
Modality	Method
Net net	End result/conclusion
Paradigm	Model
Push the envelope	Test the restrictions
Resource constrained	Not enough people/money
Resultful	Gets results
Right-sizing	Cutting excess
Scope down	Examine closely

Scope out	Take an in-depth look
Skill set	Skills
Solution set	Solution
Suboptimal	Not the best
Workshopping	Work on

AVOID "TECHNOBABBLE"

"Technobabble" is defined as technical language that goes over the head of anyone who is not well versed in it—whatever *it* may be. And, with today's advancing technological industries, you have to be considerate, with both your words and your writing, of people who are not directly involved in your specific field. One executive in the computer consulting business says:

> We make a conscious effort to cut back on the computer jargon when writing to our clients. Using jargon only makes clients feel inferior and self-conscious, and it isn't necessary to make our point.

Be sure to write in language that everyone can understand. Also be sure that your writing is not misinterpreted. Try gearing the letter toward someone who is totally uninvolved in the subject about which you are writing. If they can get the gist of your message, then you have written simply and clearly. The obvious exception is when you are writing to someone in your same field. In that case, you can be as technically focused as you want to be. Remember that you are writing to impact the reader in some way, not to totally confuse them. They can do nothing about your subject if they know nothing about it.

Here is an example of techno-jargon that would baffle any non-technical person:

> HCO, Inc. will develop a real-time interface between Lexor and Rocky Mountain Corp. This interface will be bi-directional, and assumes approximately ten business event transactions will be supported (based upon current design documentation). The interface will be implemented in an asynchronous fashion, to provide greater reliability and system scalability, using an event queuing/routing solution such as Microsoft BizSpeak.

Here is the "understandable" version—in this case, it had to be a little *longer* to simplify:

> HCO, Inc. will develop software that will allow Lexor and Rocky Mountain Corp. to communicate with each other instantly. Data can be transferred from Lexor to Rocky Mountain Corp., and vice versa. It will be transferred between the two systems in ten separate distinct events, which will be triggered by end-users in either system. When the data transfer process is triggered, the information will be placed in a queue for execution. The queuing of data transfer requests

will allow for greater system reliability, and will also make it easier to grow the system, store more data and/or add new features. Microsoft BizSpeak will be used to support the queuing process.

AVOID "BUREAUCRATESE" AND "LEGALESE"

Another way you can simplify your writing is by avoiding bureaucratese and legalese. Their respective definitions are self-explanatory. Here is an example of legalese that makes the paragraph unclear, unpersuasive, and passive:

> In balancing the interests of the party of the first part with the party of the second part, full factual development from the facts herein is needed in order to ensure an equitable administration of justice.

Here, we get more information from fewer words, and with a more active tone:

> The court needs full factual development in order to ensure a fair administration of justice for both parties.

Another legalese example . . . .

> A duty of care to the herein above mentioned plaintiff was breached by the defendant when the slippery floor was left unmopped by the defendant.

The clear alternative . . . .

> The defendant breached her duty of care to the plaintiff when she failed to mop the slippery floor.

DON'T USE NEEDLESS DETAILS

While it is necessary to provide supporting details—facts and statistics—when you are making a recommendation, or drawing up a proposal, be sure to present *only* what the reader needs. There is nothing worse than getting a document loaded with numbers or obscure terms that mean nothing to you (remember to put yourself in the reader's shoes). So, wade through your information carefully, and be sure that you are sending exactly what is needed. If you have statistics to share that are important to the reader, then reference them, and attach them to your brief document. Grant your reader the chance to go over the details at his or her leisure.

LIFELESS VERSUS LIVELY WRITING

Business is competitive today—globally competitive. Of course, you want to stand out with prose that is simple and lively, and with words that demand action and attention. So, use language that is alive and kicking, and replace the lifeless phrases of the past:

Detailed and Lifeless	Simple and Alive
Along this line	So
At hand	Here
Attached herewith	Attached
Avail yourself to the opportunity	Take the chance/try
We wish to state	*Just state your case
We hereby advise	We advise/suggest
I solicit your kind indulgence	*Just ask
Under separate cover	Attached
With your kind permission	*Just ask
This is to inform you that	*Just tell it
I wish to call your attention to	*Just state it
I am writing to tell you that	*Just tell it

SUMMARY

The theme of this section is: simplicity. Keeping your writing simple is the best way to get your point across in today's global, complex workforce, where face-to-face interaction is not always possible. So, *write* like you are face to face with someone, and *keep it simple.* Remember that your reader may only have time for your first paragraph, so make it a good one; then attach whatever detailed information is needed—if any—so your reader can look over the particulars on his own time.

Know what to avoid: *anything* that is unnecessarily complex or abstract. Avoid showy words, out of date words and phrases, buzzwords, and technical jargon. Liven up your writing with simplicity and clarity. Learn to write in a direct manner that gets results. Remember that no one has ever complained that something was too easy to read.

▶ ACCENTUATE THE POSITIVE

> Words are a lens to focus one's mind.
>
> —AYN RAND

It is impossible to overstate the power of positive focus in your writing. To write positively requires that your mind think in that same direction. So, focus on the good stuff, on what you *can* do, not what you can't; and stay far away from negative thinking and a negative tone. Realize that you paint a picture with your words, and that it is up to you to make it good enough to hang in the Louvre.

TONE

The tone you use in your writing plays a vital role in your ultimate success with any issue. It conveys your attitude, your personality, and even how you feel about your reader. Whatever your personal style, it is important to remember that, when you are writing, you are, in effect, "talking on paper." And, in today's world, where so much business is done through writing, people are "listening." So, just what kind of tone should you use in your letters?

The answer is simple: Be nice, and you will never regret it—you are also much more likely to get what you want. Be unpleasant or insulting, and you can be assured that your letter will get about as far as the recipient's trash can. For example, if you write in a terse, abrupt manner, then you come across as a brusque, unfriendly person. If you write in a flowery, impish way, your reader may think he can walk all over you. But, if you write like a warm, sincere, supportive professional, then that is how your reader will see you. Having a respectful and kind tone in your writing is the best way to boost your chances for winning business relationships. Remember: "You reap what you sow."

Negative, boorish tone: Here is a memo from a manager to his employees regarding an infraction in the dress code by one of his employees:

To: All Employees
From: Russ Yates

Date: Oct. 1, 20XX
Re: Dress Code

For women: Effective immediately, there will be no more short skirts allowed in the office. This means anything above the mid-calf. Please review the dress code rules in your new hire packet. You may only wear skirts that are two inches above the knee. If you are caught wearing a short skirt, you will be sent home (unpaid) immediately.

For men: Effective immediately, there will be no more jeans or sneakers allowed in the office. Please review the dress code rules in your new hire packet. Only wear dress slacks and shoes to work. A total disregard of these rules will mean that you will be sent home immediately and you will lose your pay for the day.

The above memo is condescending and challenging. It uses a negative term like "caught," which immediately puts people on the defensive. It also not only puts off the employees who do abide by the dress code, but it demonstrates the manager's clear lack of grace and warmth. Subordinates will not follow such an uncouth leader for long.

Positive, respectful tone: Here is how that manager can improve his writing style:

To: All Employees
From: Russ Yates

Date: Oct. 1, 20XX
Re: Dress Code

Just a reminder about dress code regulations for men and women: Women: Please keep all skirts to two inches above the knee, no shorter. Men: Please refrain from wearing jeans and sneakers to the office. Take a quick minute to review company dress code policy—it is on pages 12–13 in your new hire packet. You may all dress comfortably for dress down Fridays. As always, my door remains open for any questions or concerns.

Thanks for the continued great work—keep it up!

The readers—even the one who broke the dress code rule—leave this letter with a much more positive feeling. They are reminded that the boss is paying attention to everything and is also there for them. There is no need to be threatening, as consequences for another infraction are already listed in the referenced new hire packet, and are imposed individually anyway.

Clients, colleagues, customers, subordinates, and managers all like to deal with people who are open and gracious. Even when you are the person in charge, remember that you are not running the show alone, and that your people will eventually bail out in droves if you spread negativity. In fact, using a positive tone is so important that, in many cases, it is your *only* option.

HOW TO DEVELOP A PLEASING TONE

Thomas Jefferson once said: "On no question can a perfect unanimity be hoped." He was right. There is no way for everyone to agree about everything all of the time. So, the only workable option is to express yourself with dignity and poise, and afford others that same opportunity. You give people the chance to respond to you with grace when you approach them with grace. Whether you are a pleasant, positive person by nature, or you can't make it through the holidays without shouting, "Bah, humbug," here are a few tips on how to develop a pleasing tone in your letters:

- *Be natural:* This means, be yourself. Don't write anything that you wouldn't say, like "I solicit your kind indulgence," instead of, "Can you send me the report on ..." or, "I regret to inform you that ..." instead of, "We wish we could admit all of our applicants, but ...." Of course, you have to keep in mind all of the different ways you speak to people—you wouldn't use the same tone with your children as you would with your boss—and always use professional discretion.

- *Use everyday language:* Apply the "human factor." Clients can't build a relationship with a machine; so show them that you are human by avoiding terms like "modus operandi" and "sine qua non." Replace those phrases with words like "method" and "essential." Don't make the common mistake of equating stuffy and impersonal with respectful—it is not. It is simply stuffy and impersonal.

- *Don't get sloppy or careless:* Even though you should be warm and personal in your tone, be sure not to get sloppy with your prose. Remember that you *are* at work and that consistency with proper grammar and sentence structure is key to maintaining a positive professional image.

- *Be humble:* It is critical to stay in touch with your workforce if you want long-term success and mutual respect. So, use language that your employees can relate to—but don't talk down to them. Remember: they simply occupy subordinate positions to yours; they are not inferior human beings.

- *Cut out "angry" words and phrases:* They do nothing more than provoke an argument, which is not your goal. Delete anything that sounds accusatory or patronizing: *lazy, alibi,* or *blame.* Avoid libelous words such as *fraud, cheat,* or *unethical,* or you will need a lawyer before you know it. Remember that people with opinions differing from yours are not necessarily crazy or ignorant. Adopt Thomas Jefferson's quote: "I tolerate with the utmost latitude the right of others to differ from me in opinion without imputing to them criminality." There is nothing more annoying than a preachy, self-righteous tone—and it will eventually come back to bite you.

- *Emphasize what you can do, not what you can't do:* Focus on the upside of the situation, and offer alternatives if possible. Quick true story: Two families signed contracts to buy the same house. When the developer realized the error, they decided to give the house to the family whose contract had arrived in the office first. Then, the developer explained the situation to the other family (who was now out one house) and immediately offered another home at the same price, but with several upgrades. Both families accepted happily, appreciated the honesty, and ended up with their needs met.

- *No negativity:* Unless you are writing a letter that absolutely demands some negative content (like a collection letter), then avoid negative words and phrases as much as possible. These are words like: *impossible, terrible, never,* or *crisis.*

- *Do more than you have to—go above and beyond:* This means you should help someone even if you think it won't directly benefit you. Your tone has a way of opening—or closing—the most unexpected doors. Imagine you are the director of a private high school, and a parent new to your town writes to ask about your school for her seven-year-old daughter. You can either respond abruptly with a snappy, "We are a high school, not an elementary school." *Or,* you can reply with a helpful and kind, "Thank you for your interest in our school, and welcome to Bellport! We are a high school, with kids in grades nine through twelve, but we look forward to having your daughter join us as a ninth grader in a few years. Here is our recommendation for a wonderful local elementary school . . ."

- *Time your letters for best results:* Although you want to allow yourself the time to think through each letter you write, you should also be respectful of the reader's time. This means that you should respond as promptly as possible. This reflects positively on you as a caring and diligent person. For large matters, you can simply reply with an initial acknowledgment to the sender to tell him that you received the letter and you are looking over it.

- *Show enthusiasm:* Let's face it, eagerness and passion are contagious. People usually react well to a positive outlook. So, don't be afraid to use words or phrases like, *beyond compare* or *invaluable asset* to emphasize the positive.

- *Use contractions to warm it up (or **cool it down without them**):* Have you ever noticed the way newscasters and television anchors present their material? Of course their job is to present the news seriously and professionally. But they also personalize and warm up their rhetoric by using contractions when they speak. If they didn't do this, they would sound stilted, cold, and unnatural. The same goes for your writing. If you want to sound warm and accessible, try this:

> "We'd like to break this project up into a few manageable chunks before it is crunch time" sounds better than: "We would like to break this project up into a few manageable parts before it is crunch time."

> Of course, it is sometimes necessary to make your point stronger by drawing it out more, and by *not* using contractions:

"We do not recommend that you go forward with this project right now—it is only a matter of weeks before three other larger projects begin."

- *Read your letter aloud before you send it.* A good final test for your letter is to read it aloud to yourself to see how it flows. If it sounds natural and paints a professional and pleasant picture, then you are ready to send it.

BE MINDFUL OF YOUR MOOD

One thing that we sometimes forget to take into consideration with our busy schedules is what kind of mood we're in when we write something, and how that mood might affect what we say. And it is the rare person who is in a happy, lovely mood all day every day. So, you need to be careful about what you send out when you are in "a mood." You might be tired, stressed, or just plain angry at something. But, it always looks worse when you transfer those feelings to paper and send it to someone else.

One executive uses this technique when he has an "I shouldn't be dealing with humanity" day:

> If I'm on a time crunch, I try to at least have one other person quickly read my letter and edit it for hostility. But, if I have more time, or the issue is a big one, I write and revise it, and even ask for my boss's opinion, and then proof it again before sending it.
>
> Too many people today are afraid of using the expert resources—their supervisors or more experienced colleagues—that are right there in front of them, just waiting to be a mentor to someone. I think it shows that you are knowledgeable, mature, professional, and focused when you ask for a second opinion. You grow quicker when you ask questions.

Don't be too ready to show others they are wrong. Remember that you are writing to change something, not to vent your feelings. You can even try to validate both points of view, and then gently persuade the reader with your opinion. And it is fine to vent your feelings privately (writing them down does help), but never send a letter in anger. Give your rage a chance to dissolve a bit, and then write something that is productive, rather than sniping:

> *Sniping:* "Dave, what were you thinking with that proposal? I'm stumped by your stupidity. Redo it so it looks like something we can actually work with."
>
> *Productive:* "Dave, let's meet and go over that proposal. I've got a few ideas I'd like to run by you—looking forward to getting your thoughts."

Of course you are human, so you will have plenty of slip-ups. But, that also means that you will have plenty of opportunities to make amends! So, *when* the time comes, be prepared to write a letter of apology. You don't have to stumble all over yourself in total reverence to your reader; you just need to maintain some sense of compassion and humanity with your words. You will be proud of your professional self in the end.

SUMMARY

Tone is a powerful and important tool in all business communications. You can bring grace and dignity to any business letter by using your words to paint a positive, beautiful masterpiece. You do this by writing in a sincere, professional, optimistic manner, regardless of the subject content or the receiver's demeanor. This theory follows the age-old admonition from the Bible: "Do unto others as you would have them do unto you." In other words: If you honor people with your words, you honor yourself in the process.

Emphasizing the positive is as easy as saying: "Always remember me," instead of "Never forget me." But there are some time-tested techniques that can help you develop a pleasing tone:

- Be natural, not stiff.
- Use common language, not stilted phrases.
- Don't be sloppy.
- Be humble, not haughty.
- Remove "angry" words such as "lazy" or "blame."
- Emphasize what you *can* do.
- Avoid negative, demoralizing words.
- Do more than you have to—help someone, even if you think it won't affect you directly.
- Respond promptly to people, unless the topic is one that needs time to mull over—then let them follow your thinking process.
- Show enthusiasm—it is contagious.
- Use contractions to warm it up; don't use them if you want to emphasize something.
- Read your letter aloud before sending it—that puts you in the reader's shoes and gives a great overall image that your words present.

It is also important to pay attention to your mood when you write. If possible, you might need to wait a few days until your disposition improves, so you can write with clarity and professional candor. You want to avoid hostile words at all costs, as they only serve to worsen a situation. Remember that you are writing to persuade, not to vent your anger. You have to be at your best in your letters, even if you are not in the best mood when you write them. Ultimately, you have done your job well if you have written to accentuate the positive and you leave circumstances better than you found them.

► AVOID CERTAIN WORDS

> Remember that we often repent of what we have said, but never, never of that which we have not.
>
> —THOMAS JEFFERSON

We have established by now that the first way to lose your audience is to use words that fly over their heads, bore them to tears, or maybe even offend them. In today's competitive market, you can't afford to waste one moment, one stroke of your pen, or one tap on your keyboard on words that don't work. This section is dedicated to helping you wade through commonly confused words, so you will be able to use the right words at the right time. It will also cover words and phrases you should avoid, such as cliches and slang. And it will provide a fresh outlook on "sensitive" words. You need to be aware of the words that will hinder your efforts, so that you can focus on those that will ensure your success. More on these topics follows in Chapter 4: Grammar.

Commonly Confused Word	Definition
Accept	recognize
Except	excluding
Access	means of approaching
Excess	extra
Adapt	to adjust
Adopt	to take as one's own
Affect	to influence
Effect (noun)	result
Effect (verb)	to bring about
All ready	totally prepared
Already	by this time
Allude	make indirect reference to
Elude	evade
Illusion	unreal appearance
Altar	a sacred table
Alter	to change

Among	in the middle of several
Between	in an interval separating (two)
Appraise	to establish value
Apprise	to inform
Assure	to make certain (assure someone)
Ensure	to make certain
Insure	to make certain (financial value)
Beside	next to
Besides	in addition to
Bibliography	list of writings
Biography	a life story
Breath	respiration
Breathe	to inhale and exhale
Breadth	width
Capital (noun)	money
Capital (adjective)	most important
Capitol	government building
Cannot	the word cannot is one word—it should never be spelled *can not*
Complement	match
Compliment	praise
Continual	constantly
Continuous	uninterrupted
Decent	well-mannered
Descent	decline, fall
Disburse	to pay
Disperse	to spread out

Disinterested	no strong opinion either way
Uninterested	don't care
Elicit	to stir up
Illicit	illegal
Eminent	well known
Imminent	pending
Envelop	surround
Envelope	paper wrapping for a letter
Farther	beyond
Further	additional
Immigrate	enter a new country
Emigrate	leave a country
Imply	hint, suggest
Infer	assume, deduce
Irregardless	*Not a word—the word is "regardless"
Its	Possessive pronoun
It's	Contraction for it is
Loose	not tight
Lose	unable to find
May be	something may possibly be
Maybe	perhaps
Nuclear	The word is pronounced "nuclear," not nuke-ular
Overdo	do too much
Overdue	late
Persecute	to mistreat
Prosecute	to take legal action

Personal	individual
Personnel	employees
Precede	go before
Proceed	continue
Proceeds (noun)	profits
Principal (adjective)	main
Principal (noun)	person in charge
Principle	standard
Realtor	The word is pronounced "*real*tor," not rea-*lit*-or
Stationary	still
Stationery	writing material
Supposedly	The word is pronounced "suppos*ed*ly," not suppos*ably*
Than	in contrast to
Then	next
Their	belonging to them
There	in a place
They're	*they are*
To	on the way to
Too	also
Weather	climate
Whether	if
Who	substitute for he, she, or they
Whom	substitute for him, her, or them
Your	belonging to you
You're	*you are*

AVOID CLICHES

Cliches are overused expressions. They become overused because they seem to describe certain situations in a light, colorful, and very truthful way. The problem with using them is that, once they become overused, they start to sound like a fan running softly in the background, and they lose their effectiveness. They just don't deliver your message with the same force as your own original words.

There are hundreds of cliches, but here are some of the most common ones found in business writing:

Add insult to injury	Hit the nail on the head
Back to the drawing board	If worse comes to worst
Ballpark figure	In a nutshell
Beat a dead horse	Interface
Behind the eight-ball	Knuckle under
Beside the point	Last but not least
Bottom line	Lesser of the two evils
Business as usual	Letter perfect
Clear the air	Low man on the totem pole
Cream of the crop	Make ends meet
Dialogue	Mark my words
Dog-eat-dog	Meaningful
Do's and don'ts	Meet your needs
Dot the i's and cross the t's	More or less
Eleventh hour	Movers and shakers
Few and far between	Pack it in
First and foremost	Pay the piper
Get a leg up on	Point in time
Grin and bear it	Rat race
Hand in glove	Roll with the punches
Hands on	Run it up the flagpole
Handwriting on the wall	Spill the beans
Heads will roll	State of the art
Hem and haw	Take the ball and run with it
Hence	Too many irons in the fire
Heretofore	Well and good
Hit pay dirt	

The idea with cliches is that any one of them can be replaced with solid information. Sometimes, we can't help ourselves, and we use them because they describe a situation "to a tee," or convey a thought in a light, humorous manner. But, we tend to use cliches when we either don't know the information we're referring to or when we're unsure how to word something. So, take a chance and be original! It shows you know your stuff—and, at the very least, you will be heard.

■

WORDS OF WISDOM
"Use correct English; avoid slang. Instead of, "It is like tomorrow is a new day," write "Tomorrow is a new day." Don't say, "I went tomorrow is a new day, and he goes, 'No, tomorrow is not a new day.'" Use the verb, *said* instead of *went* or *goes*."

—JIM WHITTLESEY,
EXECUTIVE VICE PRESIDENT
OF WHITTLESEY MANAGEMENT SERVICES

■

SLANG

Slang is defined as "nonstandard" terms—not vulgarisms—that are used in the conversational style of a given culture. Standard English is the standardized, well-known language structure. It is important to understand that the recommended conversational writing style of today does not translate into a free pass to use slang. What it *does* mean is that, while most slang terms can be found in the dictionary, they are not suggested terms for business correspondence—even the most casual kind.

For one thing, slang is vague. It does little more than just give the reader an overall picture of a situation, rather than provide the descriptive words needed to clarify something. It also demonstrates lack of ability to come up with your own words. The reader might be left wondering about your intelligence. It is also unprofessional—there is no other way to categorize it. Use standard English, with a warm and conversational tone, and you can never go wrong.

Here are a few slang terms to avoid:

Axed
Bogus
Booted
Break a leg
Bummer
Canned
Don't blow it

Don't drag it out
Get a grip
Get the lead out
Get with the program
Green light
Hosed
Keep your shirt on
Out of whack
Thumbs down
Thumbs up
Up to snuff

BIASED WORDS: BE AWARE, BUT SENSIBLE

These days, it seems like there's very little you can say or write anymore without getting yourself into some kind of trouble by using the "wrong words." A new and increasingly diverse workforce inevitably brings change, and with it, an obvious need for *some* degree of sensitivity to that change. The question then remains: What are the words we need to look out for, and how carried away are we going to get with this stuff?

Biased words are words that are considered discriminatory, or degrading, to particular groups of people. Three main areas of concern in the workplace are: gender, race, and disability issues. The best rule of thumb is to *always* be aware of your audience, and *always* use your common sense. Since you cannot possibly control what is going to offend every person you meet, you will do well to simply be aware and use your common sense. Following are some tips on how to handle sensitive words:

GENDER ISSUES

This is a pretty easy one: just use the person's title; or, use "person" at the end of certain terms:

Gender-specific	Job Title
Ad man	Advertising executive
Fireman	Firefighter
Housewife	Homemaker (*or* stay-at-home mom or dad)
Mailman	Mail carrier
Policeman	Police officer
Salesman	Salesperson
Spokesman	Spokesperson
T.V. anchorman	T.V. anchor
Weatherman	Meteorologist

When referring to a couple, don't make any assumptions:

No: Mr. Ammerman and Caryn
Yes: Mr. Ammerman and Ms. Fetzer

Use professional, rather than personal descriptive terms:

No: Robin Benoit, a lovely associate
Yes: Robin Benoit, an experienced associate

RACE ISSUES

The biggest lesson here is to focus on the *person*, not on the race to which they happen to belong. Avoid stereotyping by simply emphasizing the person's individual, professional characteristics and qualifications, not racial characteristics. A person's race is irrelevant to their level of intelligence and to their job performance. The only exception to this is when a person is filling out a personal profile in a human resources department. That is done expressly to help the Census Bureau with their workforce statistics. So, just completely omit any and all racial references. They don't belong in business, nor do they belong anywhere else. Race is, in effect, a non-issue in your business writing.

DISABILITY ISSUES

The same rules apply with disability issues as apply to race issues. A person's disability is a non-issue where their professional qualifications are concerned. Of course, there are some obvious physical and mental limitations where certain people are concerned, but those are issues that are between a supervisor and an employee. They have no bearing on what you should or should not write in a business letter. Address the *person*, not their disability.

SUMMARY

The message of this section is clear: In all of your writing, be accurate, be original, be professional, and be sensible. Study the lists of words and terms that take away from your writing effectiveness. Learn the meanings of words and how to spell them. Remember that spell check won't catch proper words that are used in the wrong context. Also, do not use words that are not words. Try to avoid cliches and apply some of your own brilliant thoughts. And, of course, use sensitivity and reason when you address someone.

Your job is to be aware of the growing and changing global workforce, and to apply your best thoughts to that process. Try to maintain professionalism and understanding in all that you write.

▶ Add a Personal Touch

> Be yourself—let your business writing represent your personality.
>
> —Jim Whittlesey, Executive Vice President

The message is simple: Your writing is you. It shows how your mind works. In fact, many of your business contacts may only know you through your writing, so be aware of the written image portrayed through your business correspondence. One elementary school teacher takes this theory a step further, saying that she would rather stand naked in front of a room full of parents than have her writing skills exposed. If you approach your writing with the knowledge that you can—and *should*—be yourself, then you are on your way to success!

YOUR STYLE

Style: It is not what you say, but *how* you say it. How true. E. B. White (co-author of the acclaimed *Elements of Style*) asserted that all writers reveal something of their spirit, their habits, their biases and their abilities by the way they use their language. And, when you write, you are making not only a first impression, but also a permanent impression on the reader. So, make it a good one.

We have established the fact that a conversational tone is the trend in business writing today. Our letters come across warmer, livelier, and easier to understand when we write naturally. So, let your personality gracefully flow through your words and watch your writing style unfold before you.

It is first important to distinguish between a business-appropriate style, and one that may be you but is not good business form. An executive for a technical company says this about one of his employees:

> He is a brilliant analyst, but he is from another country and only speaks English sparingly. Unfortunately, the first words he has learned are a few basics and a few expletives. So, whenever he writes, he doesn't necessarily realize how he's coming across to his audience. He sees his writing as his personal style, but what he's really laying the groundwork for is either a lawsuit or a client bailout. It can be funny internally, but when you imagine yourself sitting at the other end of the line, reading this thing written by this person whose personality you are not familiar with, it does not conjure up the best mental picture.

Of course, the lesson here is to be careful with your *choice* of style. Take the basics of your personality and apply them in a professional way at all times.

HOW TO DEVELOP YOUR STYLE

- *Practice some personal "Feng Shui":* Take time to get to know yourself. Learn the things about your personality that others appreciate. Are you funny? Are you kind? Are you personable and engaging? Emphasize the positive things about *you* in your writing style, and work them into your words to create a classic, original piece.

- *Ask yourself some key questions:* What kinds of things would you like people to say about you, based on your writing? What kind of qualities would you like them to see in you? What kind of business relationships are you looking for? What level of professionalism are you after in your career?

- *Use "I" and "You":* Don't be afraid to refer to yourself as "I" in your letters. If you try too hard to take the focus off yourself, then your writing comes off sounding unnatural and stilted. Better to let it flow easily and normally, and sound like an actual human being. And, it is in keeping with the widely accepted conversational tone of today. You can also use the word "you." But, be careful not to overdo the "you" approach, or you will sound patronizing:

  A bit patronizing: Your work on the Letterman project was insightful, thorough, and right on target, Melanie.
  Better: I think the whole office benefited from your work on the Letterman project, Melanie. It was insightful, thorough, and right on target.

- *Use "We" and "I":* Use "we" when you want to speak for your whole company, and use "I" when you are just speaking for yourself. As a rule, if you are in constant contact with a particular employee from another company, it is safe to stay with "I." But, if you need to convey a message that has the backing and the clout of your entire corporation, then use "we."

- *Always use discretion:* There are many different situations in business that require you to handle yourself diplomatically. When things go wrong, you can choose to either handle an issue emotionally, or you can handle it professionally. Of course, you are not expected to become a robot when you enter your office, but you should try to maintain a calm and cool manner at all times—especially in your writing. One of the best ways to express your dissatisfaction at something (or some*one*) is to turn your anger into disbelief, with a professional tone:

  Emotional: I am livid at you and the other ignoramuses in your department who dropped the ball on the Kelly account. You ought to be ashamed of yourselves. Don't ever expect to work with us again.

Professional: We are confused as to how the Kelly account fell through. Let's arrange a meeting between our two departments so we reach an understanding of what happened, and devise a plan about where to go from here.

- *Think like a CEO and avoid platitudes:* You don't have time for them. Impress your reader with the "real you," and be someone who is focused on getting things accomplished. Refer to the section *Keep it Simple* (on page 23) and avoid flashy words.
- *Apply the "Conversation Test":* Do you sound like yourself in your words? Read your letter aloud and ask yourself if your written words sound like you are actually speaking to that person. Again, this doesn't mean get sloppy; it means get real. If you sound like the real—professional—you, then you have done it!

ACTIVE VS. PASSIVE VOICE

Active and passive voices in writing set the tone in every sentence. The *active voice* directly connects the action with the person who is performing that action. The *passive voice* renders the *doer* of the action less obvious, if that person is ever identified at all. The active voice is concise and energetic, and it is the preferred writing style:

Active Voice: We recommend you file a claim.
Passive Voice: It is recommended that you file a claim.

Active voice: Let's meet soon.
Passive Voice: A meeting should be held as soon as possible.

Even though the active voice is more straightforward, there are times when the passive voice is necessary:

1. When you don't know who the subject is:
   Our proposal was submitted late because critical details were still missing.
2. When you want to emphasize the receiver:
   Hannah was accepted at Harvard Medical School.
3. When you want to put some variation into your text, or smooth thought transition:
   This year's Holiday Party will be held at Snoqualmie Falls ski lodge. It should be a warm and festive celebration—see you there!

SUMMARY

The verdict is in: authentic, original writing is the best way to personalize your business correspondence. It establishes your personal style and adds flair to what might be an otherwise dull business letter. So tap into yourself!

Find out what your strengths are and use them in every sentence you write. Discover where you want to go with your writing and how you want to be perceived by asking yourself some important questions: What kind of qualities do I want others to see in me through my writing? What kind of business relationships am I interested in? And, don't be afraid to "get personal" by using words like, "I," "we," and "you"—it sounds more natural. Also, don't get so carried away with your conversational tone when dealing with an adversarial situation—always practice discretion and professionalism.

Stay basic, simple and natural in your style. You don't have time for anything else. This doesn't mean that you are not allowed a complex thought, but it does mean that you should think in terms of Einstein's simplified brilliance: $E=MC^2$. Come up with your own style, and simplify. Then allow yourself to shine through your writing! It is a powerful and wonderful expression of who you are.

▶ TAILOR YOUR CONTENT

> "Get a good idea and stay with it. Dog it, and work at it until it is done right."
>
> —WALT DISNEY

Walt Disney built an empire that is founded on *a mouse*. If he can take an idea like that and turn it into the success story that it is today, then you can surely take your thoughts and ideas and create a beautifully written masterpiece. You simply get your ideas down on paper, then arrange them—and rearrange them—until they sing! It may not be easy, but it *is* possible.

One advertising executive learned the important lesson of editing—the hard way:

"I was doing an ad for a baseball team that seemed simple enough. How hard could it be? It was one simple sentence. I was so happy with my quick, catchy sentence that I paid no attention to the typo that I had made until it was too late and had already gone to press. The ad read:

'Rangers strive for competence in the *pubic* arena.'

Of course, it was *supposed* to read:

'Rangers strive for competence in the *public* arena.'

It is funny looking back on it now, but my company wasn't laughing at the cost of my mistake at the time. I now edit and proofread everything before I submit it, and then I edit it again ten times before the final cut."

EDIT YOURSELF

Editing can be made easier by following some basic steps and by thinking your way through the process. So, once you have your letter written, edit it for content, style, grammar, everything! Use these "Basic Training" steps to guide you through the editing process:

- **Organize your thoughts**—Have you thought everything through? Does your letter follow a logical path? Did you write with the end in mind?
- **Analyze your audience**—Have you studied and thought about your reader? Is your writing geared directly toward your reader? Have you left the door open for the reader to respond? If so, have you geared your words to elicit the response you hoped for from your reader?
- **Be clear and concise**—. . . let every word tell. Did you get to the point? Did you *stay* with the point? Have you included all necessary supporting "evidence" and/or documents? Is your message clear? Are your facts straight? Have you chiseled away all needless words to be sure you are not rambling?
- **Keep it simple**—Are you focused and brief? Are you forgetting any critical information in your quest to simplify? Is it easy to understand?
- **Accentuate the positive**—Are you emphasizing the "good stuff?" Have you omitted all negativity and anger? Have you maintained a personal, yet professional tone?
- **Avoid Certain Words**—Have you stricken all words that hinder your success? Have you created any of your own words that do not exist?
- **Add a Personal Touch**—Is your personal "signature" in every sentence? Do you have a style that shines through your message like a lighthouse in a storm?
- **Tailor the Letter**—Have you written, rewritten, edited, and then *re*-edited? Do you have the opinion of a colleague or friend?

Once you have followed these steps, take a mental break—at least a few hours, but preferably overnight. Then go back over what you have written *one more time*. You can look at it with fresh eyes and a clear mind, and any mistakes will pop out at you like flashing Las Vegas lights.

After you have corrected your mistakes, ask a trusted friend, colleague or family member to give you *their* opinion. Be sure to be gracious about their thoughts even if you disagree. Remember that a big part of being a leader is being a learner—and you can learn something from everyone. Keep in mind the wise words of best-selling author James Michener: "I may not be the world's best writer, but I am the world's best *re*-writer."

Composition

▶ USE CORRECT FORMATS

> "It is easier to do a job right than to explain why you didn't."
>
> —PRESIDENT MARTIN VAN BUREN, 1782–1862

That kind of thinking is the stuff presidents are made of—knowing how important it is to do your best job the *first* time, rather than worrying about the aftermath and damage control that comes from careless work. In order to do a job well, you need the right materials and the right tools. Think of the story of *The Three Little Pigs*, where the little pig with a house built of bricks stood the test of time and resistance because he used the best tools for the job. If you approach your business writing with that same thinking, you will not have to worry about explaining away your mistakes. You will, instead, get your intended message across to your reader by using an effective business writing tool called *format.*

The format of your letter or memo is the layout and design—and it *does* matter. Not only is using the proper format a good idea, but it is the only acceptable and professional way to write in business. The overall appearance of your document is the first thing your reader notices, before they ever even read one word. If the overall appearance looks sloppy and unprofessional, then so do you. So, remember: no matter how well written your letter is, the reader's first overall impression will be the lasting one.

If you want to make an excellent first impression, then show your true professionalism! Think of it as dressing appropriately for a business meeting, and adhere to the old adage: Dress poorly, and they notice the dress; dress well, and they notice *you*. In business writing, quality content should be what catches your reader's eye, not poor format. So, stick with proper format, and get noticed!

■

WORDS OF WISDOM
Be creative in thought,
but practical in application.

■

▶ BUSINESS LETTER FORMATS

A word fitly spoken is like apples of gold in pictures of silver.
—PROVERBS 25:11

As we have already established, treating your business writing like it is an opportunity to create a priceless masterpiece is the key to your success and enjoyment of writing. And, although formatting and the practical application of your "artwork" are not exactly ethereal and romantic, they are essential tools to successful business writing.

Luckily for all of us, there are half a dozen business letter formats from which to choose. So, unless your company has a specific, pre-determined format that you must use for all of your correspondence, you can use any of the following letter styles:

Traditional Letter Format: The subject line is two lines below the salutation. The body begins two lines below the salutation or subject line. The first line of each paragraph is indented five spaces to the right of the left margin. The signature goes four lines below the complimentary close. The company signature line is four lines below the complimentary close.

1505 Main Street
Bellingham, WA 98225

April 8, 2002

St. Joseph School
Attention: Ellen Kwader, Principal
203 Windsong Court
Bellingham, WA 98225

Dear Mrs. Kwader:

RE: Classroom Cleaning

This week custodians will be checking classrooms to be sure the correct amount of furniture is in each room. This is a very time consuming task, as furniture is often moved around from room to room. Custodians then need to reorganize the furniture.

The night custodians have been working on special cleaning of classrooms, which should be completed this evening. I have continued working with the Registrar's office to try to head off conflicts.

I met with a company who does floor matting today, and I would like to order entry mats with either our logo or school name. I will receive a proposal sometime next week, so I'll have a chance then to see how much this will all cost. I will let you know when I receive the proposal.

Sincerely,

Michelle Giron

Michelle Giron
Supplies Manager

MG/hk

Semi-Block Format: The subject line is two lines below the salutation. This format uses three space indentations for the first line of each paragraph. The signature goes four lines below the complimentary close. The company signature line is four lines below the complimentary close.

2506 Hempstead Turnpike
Glen Cove, NY 11542

December 12, 20XX

Mr. Leonard Kline
Bright Lights Electric Audits
133 Main Street
Bellevue, NY 11111

Dear Mr. Kline:

RE: Refund Application

   Thank you for all the documentation sent to me with your letter dated July 17, 20XX. I now have documentation for water and sewer credits and a refund application. Let me detail the facts in this situation from the documentation you sent, as well as from other information I have.

   The copies of the documentation you sent indicate canceled charges for the periods in 1989 through 1998. Apparently, on September 7, 1988, you completed an application with the city of New York for a "Request for Refund" for a total of $22,586.13. This corresponds to the seven (7) copies of the canceled charges from the documentation you submitted to me. From all the information submitted to me by you and by my own company, it is obvious that a credit to our account or a check was never received.

   Your letters tell us that you want to work with us in resolving past issues, and that you want to go forward in assisting us with our water and sewer audits. We are hoping that we can resolve this issue with you as soon as possible.

Sincerely,

Bruce Wills

Bruce Wills
Purchasing Director

BW/ck

Block Format: The subject line is two lines below the salutation. The body begins two lines below the salutation or subject line. The signature goes four lines below the complimentary close. The company signature line is four lines below the complimentary close.

---

616 Second Avenue
Hartford, CT 06103

October 31, 20XX

Jack Kelly
Creative Change, Inc.
Internet Business Development
55 5th Street
Newtown, CT 06470

Dear Jack:

RE: Photos Selection

I wanted to let you know that I made several content changes to the photo site. I am working with Madeline Fetzer, our artist, to add the new graphic pieces to the site this week. However, before she and I go much further with the photos, we were wondering if you could help us by creating some simple captions.

Please go to www.wellconnected.org/photos. There you will find the photos that Madeline and I selected for inclusion in your site. As soon as you have a chance, please review each one to see if you can come up with a brief caption that we can use. You can e-mail me the captions, but please be sure to attach the photo number to each one.

I will let you know when I receive your e-mails and will keep you posted about our progress. Thanks so much for your consideration.

Sincerely,

Olivia Fetzer

Olivia Fetzer
CEO, Internet Business Development

OF/ck

Full-block Format: Very similar to the Block Format, except the dateline, the closing, and company signature line begin at the left margin. The subject line is two lines below the salutation. The signature goes four lines below the complimentary close. The company signature line is four lines below the complimentary close.

16003 JFK Boulevard
Philadelphia, Pennsylvania 19102

January 23, 20XX

Joe Grisham
Work & Company, Inc.
123 Labor Drive
North Lake, PA 18040

Dear Joe:

RE: Work bids

On the following pages, please find my bid for the Work & Company jobs, as well as print bids from one of my printers. I broke the bid down into two sections, the Stationery and the Collateral Kit. I am still waiting on print bids from another printer, but these will give you a general idea of print costs.

While reviewing your materials to come up with the bid, I realized that the only thing that is going to take any time to create is the brochure. Everything else is pretty straightforward. But, based on our discussion of two versus one sample, I decided that it would not be worth it to give you only one. Both will fit the look of your existing collateral, and both will carry the same message, but each is unique in different ways. One sample will be based on the separate folder and the brochure, and the other one will be based on the integrated folder/brochure concept.

Also, to let you know, I tried to keep the bid cost sensitive, as I believe that we can work out the contents of the brochure prior to the design. Please call me with any questions or comments. I look forward to your thoughts!

Yours truly,

Cade Benjamin

Cade Benjamin
Benjamin Brown Design
CB/dk

Square-Block Format: The dateline is on the same line as the first line of the inside address, but at the right margin. The subject line is two lines below the salutation. The signature goes four lines below the complimentary close. The company signature line is four lines below the complimentary close. The sender's and typist's initials are on the same line as the company signature line.

1235 Market Street
San Francisco, California 94102

Donald Poppins   March 1, 20XX
WebWorld
341 Brick Street
Napa, CA 94558

Dear Donald:

RE: Prototype site

Thank you for giving us the opportunity to review your client's new business venture. Based on our experience and expertise in providing effective Internet solutions, I believe we can help you build a sound prototype site for YourAds.com.

Generally, we prefer to meet face-to-face with prospective clients in advance of preparing a proposal for them. We have found that interaction and feedback from such meetings gives us a much better sense of project parameters. But, given the urgency of presenting this information to your client, I realize this wasn't possible. As you review this proposal, please know that I will be happy to expand on any of the concepts and ideas presented.

Feel free to contact me at 860-888-7777, or at mickey@cchange.com after you have had a chance to review this proposal. I look forward to meeting soon.

Sincerely,

Mickey Bettison

Mickey Bettison
Vice President, Internet Development

MB/ba

Simplified Format: Omit the salutation in this format. The subject line is three lines below the inside address. The body begins three lines below the inside address or subject line. The complimentary close is omitted here. The company signature is five lines below the body of the letter.

---

325 Peoria Street
Chicago, Illinois 60607

December 24, 20XX

George Starbuck
AdCopy, Inc.
707 Elm Court
Crystal Lake, IL 60012

RE: Mechanical disk printout

Blair Schiller of Final Touch asked me to forward the enclosed mechanical, which will be printed on a computer disk jacket. The reference number for this job is 149131. The disk jacket will display a sailboat emblem, and will be available in all seasonal colors as you requested.

Please note the special attention to the details of your design request. Our team of graphic artists labored over the specifics, and made sure to send you the finest quality product we have to offer. In addition, they made sure to consult with renowned mechanical design specialists so that the end result would be free of flaws. We hope you will be as pleased with the outcome as we are!

If there are any problems with the artwork, I can be reached at 718-222-3333. Otherwise, you can contact Blair Schiller for quantity, payment, and shipping instructions.

Charles Yeager
Owner, Yeager Design

CY/ck

▶ PARTS OF A BUSINESS LETTER

The arrangement of all the parts of your business letter is just as important as the format itself. Most business people are accustomed to seeing information in certain areas of a typical letter, so your letter will read much better if you use a common form.

There can be up to sixteen key parts of a letter, including the letterhead. And, although you probably won't use all of them in every letter you write, their correct placement is essential in making your document look professional:

> **Letterhead** (See **Paper and Envelopes** for more information): This is company or personal stationery.
>
> **Date:** This is the date on which the letter was *written*, not when it was typed or sent. The date is at the top of the page, at least two lines below the letterhead symbol or sender's address.
>
> **Inside Address:** Your reader's name and address appear just as they do on the envelope. This should be flush left, and at least two lines below the date. If you don't know the reader's name, use his or her professional title:

Director of Marketing
K-2 Products Corporation
8518 North Peak Boulevard
Kennebunkport, ME 04046

Dear Sir or Madam:

> **Attention Line:** This is only used when your letter is addressed to a company as a whole, but you want someone specific to handle it. It should be flush left in the inside address, and can be either above the inside address, or immediately following the company's name. Job title is not included:

Attention Ben Dickinson
Dover Institute
140 Northwind Avenue
Seattle, WA 98101

Ladies and Gentlemen:

> **Salutation** (See **Salutations and Closings** for further information): This is your first personal greeting to your reader. The general rule here is to always use your reader's name when you know it (and spell that name correctly!):

> *Dear Dr. Ammerman.*

The salutation should be flush left, and placed two lines below the inside address.

Subject Line: The subject line consists of a few words that briefly describe the content of your letter. It is not absolutely necessary to include it; in fact, it is often omitted in current business writing. But, it does serve as a courtesy to the reader. And, it is also a useful reference for you if you need to find it quickly.

> re: Automated Processing Project

As an option, you can omit Re: and simply use capital letters:

> AUTOMATED PROCESSING PROJECT

Body: This is the actual message of your letter. It begins two lines below the salutation, or subject line. Place the paragraphs flush left, or indent them (depending on format), and use single-spacing.

Closing: This is where you sign off on your letter. It is sometimes referred to as a *complimentary close* because is it designed to close the letter in a polite, professional manner. Typical closings include: *Sincerely, Very Respectfully,* and *Best Regards.* The degree of formality you should use depends on the status of your reader (See **Salutations and Closings**).

Signature (Company and Signer) Line: This is the name and job title of the person writing the letter. It should only be used when you are writing on behalf of the *company.*

Reference Initials: This references anyone involved in the preparation of the letter. There can be up to three different sets of initials: those of the person signing the letter, followed by the dictator's, if different, and then those of the typist. Reference initials are not as widely used as they once were; but, if you do use them, use all capital letters for the signer and/or dictator, and lowercase letters for the typist:

> HLK:DAK/co

Enclosure: This indicates that you have included additional paperwork in your correspondence. You can use the word, *enclosure.* It is placed two lines beneath the reference initials. By adding an *enclosure* line, you are not only providing a courtesy to your reader, but you are ensuring that the extra information you are sending does not get overlooked:

> Enclosure

Filename Notation: This references a file name, and is placed two lines beneath the last notation. You do not need to include the word *Reference* in this information:

> whittlesey.doc

Delivery Notation: This is used when your document requires special handling. It is placed two lines below the last notation.

> By United Parcel Service

Carbon Copy (cc) Line: This line tells your reader who else is being sent a copy of your letter. You can preface the information with either *cc*, or with the word, *distribution*, if the letter is being sent to more than three or four readers. It is placed flush left and is two lines below the last notation:

> cc: Dave Gravrock, Electronic Data Systems
>   Barry Brown, Dataport Corporation
>   Bruce Crile, Owens Illinois Glass Company

Postscript: Written as *P.S.* at the very bottom (flush left) of your letter. It is placed two lines below the last notation. The postscript is simply extra information that is unrelated to the main information in your letter. It should be only one brief sentence and should be followed by the sender's initials:

> P.S. Let's get together early next week to discuss the JD Enterprises merger—I've enclosed a list of great restaurants here in San Francisco for your perusal! DWT

Continuation Page: This is any page after the first page of a document. Whatever you do, do not put the word *continued* on the first page—your reader will deduce that fact when they turn to page two. Information included in this portion is as follows: the addressee's name, the date, and the page number. You should put this information at the top left corner of each page, flush left:

> Cade Benjamin
> December 12, 20XX
> Page 2

► MEMORANDUMS

"Blessed is the man, who having nothing to say, abstains from giving wordy evidence of the fact."

—GEORGE ELIOT

A memorandum (or memo) is *just* the place to abstain from wordiness! Its sole purpose is to serve as a short, informal, written business communication—to briefly outline a particular situation, transaction, or agreement. It also helps you keep track of your business dealings by providing a *paper trail*. And, although paper memos (hard copies) have become practically obsolete in the age of advanced technology, *electronic memos* are widely used. They still serve the same purpose: They are used for policy statements, informal reports, company announcements, or directives.

Memos have long been referred to as *in-house correspondence* because they are usually informal documents sent between employees who work within the same company. But, memos today are now also sent between associates, both in and out of your company. Just remember this: Wherever your memo is sent, or however informal it is, you still need to follow standard business writing guidelines.

Just as with any other form of correspondence, you should think about what you want to write before writing it—and always remember your audience. You should also watch your spelling and grammar (use those spell-check and grammar check options in your computer system if you have them, but also remember to proofread it yourself because they don't always catch everything), and be mindful of your tone. And, never forget that you are at *work— always* maintain professionalism.

A memo can be sent to just one person, or it can be distributed to a very large number of people, depending on who needs to read it. So, measure your level of familiarity in your writing by the relationship between you and your reader. Be personable, yet not presumptuous—and, as always, be careful with your words. Above all else, be clear!

MEMO CHARACTERISTICS

Memos are . . .
- *Concise*
- *Less formal than business letters*
- *In-house correspondence*—memos are written to share information between employees from the same company, or sometimes to an employee from another firm with whom you work closely
- *Not typed on company letterhead*

Memos . . .

- *Have two common formats*: paper note, standard (includes electronic)
- *Have no inside address, salutation, or complimentary closing*
- *Don't have to provide background information*—the writer doesn't usually have to explain background because the reader generally already knows the basics. The exception is report memos, which usually include some sort of background or description of a situation
- *Provide evidence of conversations and meetings*

In Memos . . .

- *There is wide use of bullets and numbers*
- *Jargon is okay* because they are for in-house use and are less formal
- *Standard writing rules apply*—follow all of the basic writing rules, such as watching tone, grammar, spelling, and layout

PARTS OF A MEMO

There are fewer parts of a memo than there are of a letter. But, with the exceptions of the heading and the body, many of the fundamentals are the same:

Heading: This is the opening of the memo. It includes the date, the name of the recipient, the name of the sender, and the subject of your memo. The key difference from a letter heading is that a memo does not include a salutation:

TO: Chris Aguilera
FROM: Gordon Smith
DATE: January 4, 2002
RE: Shareholders Meeting

Body: The body of a memo follows the same rules as the body of a letter. The only difference is that, because a memo does not include a salutation, the body starts two lines after the subject line.

Enclosure: This indicates that you have included additional paperwork in your correspondence. You can use the word *enclosure*. It is placed two lines beneath the reference initials. By adding an "enclosure" line, you are not only providing a courtesy to your reader, but you are ensuring that the extra information you are sending does not get overlooked:

Enclosure

Filename: This references a file name, and is placed two lines beneath the last notation. You do not need to include the word *Reference* in this information:

whittlesey.doc

Delivery: This is used when your document requires special handling. It is placed two lines below the last notation.

By United Parcel Service

cc: This tells your reader who else is being sent a copy of your letter. You can preface the information with either *cc,* or with the word *distribution,* if the letter is being sent to more than three or four readers. It is placed flush left, and is two lines below the last notation:

cc: Dave Gravrock, Electronic Data Systems Barry Brown, Dataport Corporation Bruce Crile, Owens Illinois Glass Company

Continuation page: This is any page after the first page of a document. Whatever you do, do not put the word *continued* on the first page—your reader will deduce that fact when they turn to page two. Information included in this portion is as follows: the addressee's name, the date, and the page number. You should put this information at the top left corner of each page, flush left:

Cade Benjamin December 12, 2001 Page 2

MEMO FORMATS

Paper Note Memo Format

This is the typical "From the desk of . . ." paper memo, and despite the onslaught of computer-generated correspondence in today's workforce, it is still widely used. This handwritten memo style is quick, personal, and effective. It serves a dual purpose: it shows your reader that you care enough to send a personal note, and it also spares both you and your reader from having to spend one more moment reading yet another e-mail. Here is a sample:

From the desk of . . .
Mike Ammerman

May 10, 2002

Olivia:

Outstanding proposal you put together.
With Brown & Benoit now on board, we can't lose!
Enjoy your trip—you have earned it.

MA

A note memo should only be a few sentences long at the most, and is also used only for very informal correspondence. You wouldn't want to handwrite an entire report for a client, for example, just to make it seem more personal. Use your good, old-fashioned computer for that.

Standard Memo Format

The standard memo format has been around for a long time, and it is still in full use. And now, with the development of computer software programs, a variety of memo templates have been made available for today's workforce. So, unless your company is very particular about the memo style you use, you can safely choose from any of the templates in your word processing program.

In a standard memo format, you can organize the body of your memo in whatever way gets your message across clearly and simply. Typical formatting tools include bullets, numbers, subheadings, or just plain text. Whatever memo format you use, be sure to include the date, a "to" and a "from," names of all people receiving a copy, and the subject for easy reference. The following is Standard Memo Format:

WESTAR COMPUTER ENTERTAINMENT

Interoffice Memo

To: Hannah Leigh
From: David Pfaff
Date: October 31, 20XX
Cc: Ben Dickinson
        Dave Stolp
        Kay Brown
        Caryn Fetzer
        Ben Ammerman

DDone SLEEVE PROJECT CHANGES

Today there was a change made to the DDone sleeve project that will affect almost everyone. We are going to incorporate a sticker on the front of the sleeve (lower right hand corner) to include a call out of some additional features of the demo disk.

Starting Wednesday the 31st we will sleeve 20,000 DDone units in San Jose, Fresno, and Carrollton (60,000 total). After that, no DDone units will be sleeved until we have the stickers needed. I will work closely with Sven regarding product holds, MSQ moves, and so on during this time.

The first 40K of disks and sleeves kitted for CTI will ship as they were built. No other kits should be built at this time. Kits for CTI going forward will include the new stickers once they arrive in San Jose. Instructions for applying these stickers will be forwarded along once we know the shape and design of the sticker. Canadian sleeves have not been produced yet, so the artwork for that sleeve will be changed—no need for stickers.

Let's commit to seamless integration of this new system today!

dp

One major difference between standard paper memos and e-mail memos is length: paper memos can be longer, while e-mail memos should be kept as short as possible to make it easier on the eyes of the reader.

MEMO REPORTS

Although the word *report* usually connotes a lengthy, detailed document, the fact is that a report can come in all shapes and sizes. It can be big or small, short or long, and it can be written in a variety of formats: letter, memo, transmittal, company fill-in forms, credit, or progress are a few examples.

Bigger, more formal reports generally include an executive summary, a statement of purpose, relevant data, formulae, tables, graphs, charts, procedures, conclusions, recommendations, and subsequent steps—anything and everything that will help your case. Reports of this magnitude should not be sent casually as an attachment in a random e-mail. Reserve these for a planned meeting, for example, where everyone can sit face to face and examine the details together.

The smaller, everyday *memo reports* that are used in a million ways every day can include some of the same elements as their larger counterparts, such as tables, charts, graphs, and so on. But, unless that kind of supporting evidence can be contained in about a two-page memo, it should either be sent as an attachment, or as a Web page reference.

TO: Mike Simon
FROM: Cade Ryan *CR*
December 12, 20XX

RE: Annual Evaluation

Dear Mike,

I have just completed the annual evaluation of our subcontractor, HCO Consulting, and would like to summarize the highlights of my report.

I focused on three areas in my evaluation: cost control, ability to meet deadlines, and quality of work. My research involved interviews with our customer service and accounting departments, numerous clients (new and existing), and finally the employees of HCO Consulting.

I found that cost control varied greatly from one account to the next. And, although we established spending limits for each client, HCO Consulting internally moved money between accounts when they felt a particular client needed additional support. However, although they did not stay within the spending limits we directed, they never took more than 5% from any one account. And, they did not exceed the total annual budget for all accounts.

After discussions with many clients, I found that HCO Consulting had a 98% rate of meeting all deadlines. There was only one deadline that was not met, but this was due to changes made by the client. My interviews with our customer service department were helpful in analyzing the quality of work performed by HCO Consulting. Most inquiries were requests for the phone numbers of points of contact at HCO Consulting. I couldn't find a single major discrepancy for the entire year.

My recommendation, Mike, is that we continue our relationship with HCO Consulting. We should give them more discretion in budgeting matters, and we should publish a roster of representatives (with phone numbers) that can be shared with clients. The bottom line is that we should be very proud of the performance of HCO Consulting over the past year.

FINAL FORMATTING TIPS AND TOOLS FOR LETTERS AND MEMOS

- *Think ahead about your reader, your message, and the best format and tools you can use to convey that message:* Choose whatever you think best fits the situation. Include only that which you think will help you get the results you are hoping for.

- *Use parallelism for consistency:* Parallelism in writing is when topics and subjects correspond to one another in a logical way. This is particularly important when you are using any type of outline format, such as bullets and numbering. The key is to word everything in a similar style. For example, the bulleted list you are reading from right now has a directive tone that is parallel in structure—every lead sentence makes a recommendation of some sort. It would not make sense to have the final bullet say: *Letters and memos can be difficult to write.*

- *Try not to overemphasize:* If you **EMPHASIZE** everything, you emphasize **nothing.** Keep in mind the overall picture your letter or memo paints. If it looks like a jumbled mess, then that's how *you* come across—and your message gets lost in a fog as thick as pea soup. So, avoid multicolored fonts, all capital letters, all italics, or any mish-mash of words that looks sloppy. So, stand back from your letter once you are finished and see how you like it. Is it professional looking? Does it flow easily? Better yet, would *you* enjoy reading it? If you can answer "yes" to all of the above, then you are ready to send it!

- *Keep your letters and memos to one page, two maximum:* Anything longer than that starts becoming a report, which is fine, if that's what it should be. It is okay to change your mind about a document while it is still in the works. So, whether you have composed a letter, memo, or report, just be sure that your finished product is accurate and edited before sending it.

- *Write a second letter or memo for separate or related issues:* One great way to keep your letters or memos short is to focus on one major topic at a time. If you need to discuss several different issues at length, then draft several different letters, each with a separate heading. Or, if you have an issue that is parallel to the one you are discussing, compose a separate letter to discuss it.

- *Don't make any changes that affect the readability of your letter:* Don't reduce the font size to where it is readable only with a magnifying glass just to keep your letter to one page. Either edit your text to fit onto one page, or make your letter two pages. Conversely, don't make the font huge just to fill space— the letter just ends up looking like you made the font huge to fill space.

- *Use attachments or Web links (URLs) for detailing information:* Anything in your memo that requires a detailed explanation should be presented as an attachment, or a Web link. You can attach as many separate documents as you need, as long as you refer to your attachment(s) in your memo. Be sure to maintain some sense of logical order in your attachments, as you don't want to send a confusing stack of documents. Use this tool to support, clarify, and provide statistical or evidentiary back-up for your memo.

- *Use boxes, tables, charts, or graphs to support your point:* Whatever you think will draw attention to your subject, or will clarify it for your audience is a good tool to use. However, you should use these sparingly, or your message will get lost in a sea of shapes and shading.
- *Choose from these commonly used formatting tools*: A descriptive subject line in bold print, a direct statement of purpose as an opener, compelling points in bold in the body of the document, bullets and numbers as needed, blank space to separate issues, italics, quotes, and underlining.

▶ ELECTRONIC CORRESPONDENCE

I think there is a world market for maybe five computers.

—THOMAS WATSON, CHAIRMAN OF IBM, 1943

Only five computers . . . that has to be the underestimation of the century. Little did Thomas Watson know what fundamental changes were about to take place that would reshape our world and change the way we do business for perhaps centuries to come. Computers and their electronic correspondence capabilities are everywhere—in offices and homes around the globe. So, it isn't hard to figure out that e-mailing is now the primary means of business communication worldwide—and, it is gaining momentum by the minute.

Electronic correspondence (or *e-mail*) has definitely become a business trend. Most companies today, regardless of the type of industry, have either already hopped on board with e-mail:

In an effort to match the demands of their growing business, a husband and wife veterinary team in Washington State is streamlining their entire business system with electronic correspondence. They will be using computers for everything from a reminder that your dog is due for a vaccine, to the establishment of a website from which dairy farmers can pull the results of their animals' health checks. Dr. Scott Kieser says, "We want everything to be as robotic as possible, so we can determine if it is the system or the people that need improvement. It is also a money-saving measure and a great simplification of our business process."

Streamlining work processes and speeding up communications are wonderful things. After all, that was the original intent with e-mail. But, as we get busier, and our electronic mailboxes get laden with "e-everything," the question now becomes: are we in control anymore? If not, we need to be!

TAKING CONTROL OF E-MAIL

"Three million e-mails are sent throughout [our company] per day. 200 is the average number of e-mails waiting in an employee's inbox. 30% of e-mails are unnecessary. [And], 2.5 is the average number of hours [per day] that employees spend managing e-mail." These are statistics for just one major technology firm—yet this firm probably represents most companies today. So, the challenge we face is to find ways to better manage our e-mail and become more effective and productive in the process.

There are several general areas of e-mail that you can focus on to become a more adept and effective e-writer: Formatting Guidelines, "Netiquette" (e-mail etiquette), and "E-Navigation" (electronic tips). If you follow these guidelines, you will save time, energy and money, the three most precious resources in business.

Formatting Guidelines

- *Follow all of the same guidelines that you learned about in Part One:* Organize your thoughts, keep it simple, be concise, be clear, accentuate the positive, analyze your audience, add a personal touch, avoid certain words, and tailor the letter. The same general rules apply in most cases.
- *Avoid logos and graphics:* These take up too much computer space. As an option, you can use different fonts, some color, or different type size for emphasis.
- *Use the subject line:* Make your subject line compelling and clear. It should describe to your reader what the content of the e-mail is in as few words as possible.
- *cc only those who need to read your e-mail:* Think before you copy 50 people on something that only 10 people need to read.
- *Use colors to organize and prioritize:* This is the exception to the "keep it black" rule (see *"Do's and Don't's"* ahead). This idea is meant to help you sort and prioritize your e-mails. For example, you can use red for "high priority" messages, and blue for "get to it within a week" type messages.
- *Drag and drop:* You can drag and drop e-mails into a planner or calendar to de-clutter your in- and outboxes.
- *Create file folders for your e-mails:* By creating file folders, you can sort your e-mails logically by category. Then, you can peruse those folders periodically and "zip" what is no longer pressing.
- *Use universal keywords:* Many organizations establish universal keywords that

are used to help qualify (and also prioritize) messages. One Human Resources Director says her company uses the following:

"HOT" = Read Immediately

"FYI" = For Your Information (no response required)

"AR" = Action Required

"WRS" = Status Report (no response required)

"ORG" = Organization Announcements (no response required)

- *Use <eom> for short e-mails:* This is where you only fill in the subject line of your e-mail, and it says something like: "Can you send me the monthly? Thanks <eom>
- *For long e-mails:* Include a summary and bold headers followed by subheaders—whatever it takes to break the information up and make it easier to read.
- *Don't use emoticons:* Reserve smiley faces and winks :=) and ;-)) for more personal correspondence. These are only appropriate when a colleague happens to be a friend. But, for the majority of your business correspondence, use words to convey your message.

NETIQUETTE

It is just as important to follow e-mail etiquette (or *netiquette*) guidelines as it is to come to work dressed appropriately for the job. Remember that the basic tenets of professionalism don't fly out the window just because you are sitting behind a computer—they are alive and well. So, follow these rules of netiquette, and handle yourself with grace and confidence:

- *Read your e-mail when you know you can act on it:* Try to limit yourself to checking your e-mail "inbox" only two times per day (three, maximum). That way, you won't spend all day reading, and no time acting on anything. Think of it this way: spending your whole day reading e-mails is tantamount to being splayed out on the mailroom floor under countless bags of letters! Not a pretty picture. And for sure, not productive.
- *Use a conversational tone:* Even complicated reports need to be understood. Write like you are speaking to your audience. Be professional, yet personable. Remember, it is best to keep sarcasm and humor out of most business writing, and that goes doubly for e-mailing, where no one can detect your body language, tone, or emotions.
- *Don't begin business relationships with e-mail:* Unless you have no other way to contact new associates, try not to open your business relationships with e-mail. In fact, in some countries, it is a faux pas, and unthinkable to start a business relationship online. So try a phone call—or, better yet, a meeting if possible. This will not only allow you to get a real feel for your colleague on the other end, but it will help to personalize your relationship faster.

- *Don't use e-mail if you need an immediate response:* Unless you are already in contact with someone who is expecting your e-mail, then call instead of e-mailing in this case. There are still times when a phone call is faster and more efficient.
- *Don't use e-mail to deliver bad news:* When you have to deliver bad news, the best and bravest thing to do is to either face the person in question, or call the person if this is not possible. Bad news includes things such as: firings, not getting a raise, not getting a promotion, or a poor performance report.
- *Encourage feedback:* This means it is okay to politely ask to be taken off a distribution list, or to request a short version (summary) of a report, rather than an entire report. At least you will open the dialogue for a mutually satisfying e-mail relationship. Remember: no harm in asking—just be courteous.
- *Keep it simple:* This rule should sound like a broken record by now. But, it is worth repeating in this section: If you try to write a novel to a coworker, he won't read it—there's no time. So, keep it short, and to the point.
- *No venting in anger:* With the simplicity and availability of e-mail, it is all too easy to fire off angry thoughts on the spot. Resist the urge. Don't do it. You can write your thoughts down somewhere if it helps you work through your issues, but don't press *send*. Instead, wait until you have cooled off, so you can actually write something logical to the party in question. You would be amazed at how much more effective logic is than anger.
- *E-mail is not private:* Federal laws do not yet prohibit many people (including your employer) from tapping into your private e-mail, especially e-mail sent on a computer owned by the employer. This may change someday, but until it does, write only that which you would feel comfortable posting on a billboard along a well-traveled stretch of freeway. That is about the only degree of privacy you can count on with e-mail.
- *Avoid gossip:* Don't gossip over e-mail. Not only is it a colossal waste of your valuable time, but again, remember that *anything* you write can come back to bite you.
- *Don't e-mail a quick note to a person sitting four feet from you:* Unless you need a paper trail, if you have a quick message for a coworker, then simply stand up and tell them that message. Just keep in mind that your corporate culture may encourage e-mail use.
- *Don't e-mail last-minute changes or cancellations:* Give at least 24 hours heads-up with e-mail. It may get to the recipient's computer port in record time, but there is no predicting when your message will be read. If you do cancel via e-mail (for instance if you need a paper trail), make a phone call to confirm your cancellation.

- *Ignore "Spam":* Spam is electronic junk mail that unfortunately clutters one too many mailboxes. Try not to be too aggravated by it, because it is probably going to be around for a while. Simply press the *delete* key on your computer and it is gone. No big deal.

E-NAVIGATION

In order to maneuver well in the fast-paced world of e-mail, you need to have a handle on terms, acronyms, and other e-topics. Even though the electronic world is ever changing, these tips will help you navigate through the basics of e-mail.

Electronic Acronyms

Because of the ever-widening use of e-mail, and now instant messaging—or *IM*—the world is getting accustomed to long distance, real-time electronic communication. And, as we delve even deeper into the age of technology, we continue to figure out ways to make it all work even faster. Enter the use of acronyms!

An *acronym* is defined as a word that is created from the first letters of a series of words—such as *NBA*, which stands for *National Basketball Association.* If a person were affiliated with the NBA, for example, and had to write or speak about it on a daily basis, they certainly wouldn't want to have to write out *National Basketball Association* every time they e-mailed a quick message. So, acronyms do have their place.

They are everywhere in electronic business correspondence—reports, letters, memos, and instant messages—and they are used to reference computer terms, names for organizations, and many other things. They obviously shorten and simplify information, and enable both the writer and the reader to get through material faster.

However, there are a few things to consider before you use an acronym:

- *Use the right acronym at the right time:* Remember that there is a difference between typing REC'D (referring to something having been received) in a quick e-mail to your company's president, and typing TTFN (ta-ta for now) to sign off in that same e-mail. Extremely casual acronyms like TTFN (see list of acronyms below) should be reserved for extremely casual personal correspondence.
- *Be sure your reader knows what you are talking about:* Always define each acronym for your reader the first time you use it—a simple parenthetical definition will suffice; after that, you can use it throughout the document for speed and ease. Of course, if you are drafting a memo to a coworker who knows exactly what you are referring to, then you do not need to spell it out.
- *Don't overuse acronyms:* Since most people don't type as fast as they talk, acronyms are used for convenience and speed. You want to be careful about overdoing it. The last thing you want is for your note to turn into a message that requires decoding.

Acronyms

AR	action required
BKM	best known method
BTW	by the way
EOB	end of business
EOM	end of message
F2F	face to face
FAQ	frequently asked questions
FOAF	friend of a friend
FWIW	for what it is worth
FYA	for your amusement
FYEO	for your eyes only
FYI	for your information
GMTA	great minds think alike
IMHO	in my humble opinion
IOW	in other words
LMK	let me know
MSGS	messages
NLT	no later than
NNTR	no need to respond
PLS	please
PRES	presentation (not to be confused with president)
QTY'S	quantities
REC'D	received
RGDS	regards
SR	status report
THX	thanks
TIA	thanks in advance
TMRW	tomorrow
TTFN	ta-ta for now
WRT	with regards to
WWW	world wide web
YR	your

Common Internet Website (Zone) Abbreviations

.com—Commercial Enterprises
.edu—Educational Institutions
.firm—Service Businesses
.gov—U.S. Government
.info—Information Service Provider

.mil—U.S. Military

.net—Networks

.nom—Individuals

.org—Non-Profit Organizations, such as the Boy Scouts of America

.web—Internet Organizations

Common E-Terms

Address Book: an electronic compilation of information about your contacts; usually includes: name, e-mail address, and other personal information (such as work address, notes, and phone numbers). It is up to you to determine how much information you want or need.

Archiving: a way to store and organize e-mails that you want to keep for possible future reference.

Attachment: an electronic file, such as a letter or spreadsheet, which you can add to an e-mail. It can be opened, viewed, and even edited by the reader.

Backup: a means to save your work, such as a network folder or diskette.

bcc: Blind carbon copy. This allows you to send e-mail to multiple recipients without having names and e-mail addresses displayed to other recipients.

cc: Carbon copy. This also enables you to copy multiple recipients on documents; but, names and addresses are displayed for all to see.

CD-ROM: Compact Disk Read Only Memory. This is a magnetic media that contains a large amount of information, but the files on the disk itself cannot be edited; they are read-only.

Compress: You compress a file when you want to reduce the size of the file for a large e-mail attachment, or for easier storage on a disk. In some programs, this is often referred to as zipping or stuffing a file (using a program called WinZip or Stuffit).

CPU: Central Processing Unit. This is the "brains" of the computer, where all of the calculations and instructions to run a program are made.

Emoticons: electronic symbols used in less formal correspondence to denote a feeling. Example: :=)

Floppy Disk: Magnetic media used to store or transport files from one computer to another.

Encryption: the ability to scramble a message so that no one but the intended reader can decipher it.

Folder: an electronic filing cabinet stored on your computer's hard drive, used to store and save electronic files.

Forward: this is sending a copy of an existing e-mail to a new reader.

Group List: a list of e-mail addresses. This enables you to group a list of multiple recipients under one title in your address book.

Hardware: the physical components of your computer, such as the keyboard, hard drive, monitor, and so on.

Icon: an electronic symbol that represents a computer software program, or an electronic file.

Inbox: the electronic file of your e-mail program where all new e-mails appear. E-mails that have been read can either be kept in the inbox, or moved to another file for better organization.

Internet: a global computer network that connects personal, business, and government users.

Modem: an electronic device that allows your computer to "talk to" other computers via a phone line.

Snail Mail: conventional method of sending correspondence via a postal service.

Software: computer programs that allow you to perform various functions on your computer.

Spam: electronic junk mail.

Virus: a computer program also known as "malicious logic," with a sole purpose of destroying or damaging software and hardware.

World Wide Web: an organized system of viewing most parts of the global Internet.

SOME FINAL DO'S AND DON'T'S

Following is a comprehensive list of tips that will answer many frequently asked questions, and will help you through almost any imaginable e-mail predicament:

DO's

Attachments: When sending attachments, the standard practice is to only include files that are less than one megabyte (1MB) in size. Files larger than that can bog down an entire e-mail server. For larger files, include the address of the file if it is kept on the company network, or the Web address if it is included on a Web page on the internet. Check with your network administrator for the size limit of attachments at your company.

DON'T's

Attachments: Don't use links to websites or files if you think the recipient may keep the files for a long time period of time. Websites and network files can change in a matter of days or weeks. So don't count on your link being usable for an indefinite period of time. If necessary, send a new message with an updated link.

Bullets: Use either numbers or symbols like an asterisk (*) or dash (-). Start each bullet with a capital letter. Indent bullets (if they don't naturally indent themselves), and use an extra line between bullets to emphasize the message.

Bullets: Don't mix complete sentences and sentence fragments. Using either one separately is okay, but be consistent. Don't make lists long—keep them short and to the point.

Font Color: Keep it black. This is the accepted color for formal business communications.

Font Color: Any other color than black usually has no place in a business letter or memo. Keep the colors in your personal notes and holiday letters. Sometimes you may want to use color for emphasis or to distinguish your reply from the original message.

Font Size: For readability, use font sizes between 10 and 12. The preferred size is 12.

Font Size: Don't use sizes less than 10 since they're too hard to read. Sizes larger than 12 can be used sparingly for titles and headers.

Font Type: Stick to the basic font types such as Arial, Courier, or Times New Roman.

Font Type: Don't use ornate or casual font types as they send a "too casual" message and can be unreadable.

Grammar and Spelling: Most e-mail programs have grammar and spelling tools as part of the software. Use these tools before sending any formal or semi-formal correspondence.

Grammar and Spelling: Don't use the grammar and spelling tools until you have completed composing the e-mail. This avoids wasting time making multiple checks.

Lower/Uppercase Text: Use the normal convention of capitalizing the first letter of the first word of a sentence, as well as proper nouns (place names, individual names, company names, etc.)

Lower/Uppercase Text: Don't use all uppercase text in your e-mail since this is harder to read than normal text and may annoy the reader. It also comes across to the reader as if you are YELLING! All lowercase text looks unprofessional as well.

Numbers: Spell out numbers one to nine and use number keys for 10 and greater.

Numbers: Don't use 1st or 2nd to denote ordinal numbers. Always spell these out as first or second.

Salutations and Closings: Include salutations and closings in most all e-mail messages, using the same rules that apply to other forms of business writing. E-mails that don't include salutations and closings can come across as cold, directive, and unprofessional.

Salutations and Closings: Don't include a salutation or closing if your e-mail is in response to an ongoing part of a multiple e-mail conversation. Replies to FYI e-mails also don't require a salutation or closing.

Editing: Read the message a second time. Use the grammar and spelling tools to avoid embarrassing mistakes. Check it several times before sending.

Editing: Don't fill out the *To* field until after you have finished writing the message and checked the grammar and spelling. There is nothing worse than an email unintentionally sent before it is finished.

▶ INTERNATIONAL CORRESPONDENCE

> Wisely and slow; they stumble that run fast.
>
> –WILLIAM SHAKESPEARE

Hundreds of languages are spoken in today's global workforce. And, when it comes to communicating with people who speak a different language, or who are from a different culture, the first thing you should do is . . . be wise. Take it slow and keep it simple. It is the old "tortoise and the hare" story . . . where the steady, plodding tortoise wins in the end. The rabbit may pass the tortoise along the way, but, in its haste to reach the finish line, the rabbit loses steam and ultimately loses the race. Your job is to be smart, take it easy, and stand back for a moment to *learn* about your audience. Be the wise old tortoise, and win!

The Japanese have taken this notion a step further by turning it into a successful business management principle: TQM (Total Quality Management). The fundamentals of TQM are: defining goals, long-term success, quality, getting it right the first time, evaluating performance, and striving toward improvement. If you apply these principles to your writing in general, especially where other cultures are concerned, you will enjoy a mutually beneficial relationship and a reputation as a true professional. All you have to do is be aware of everything around you, and be sensitive to the intricacies of another person or culture.

It is also important to be respectful and not condescending in your tone. We have all seen situations where people assume something about another person, act on that assumption, and then end up making a fool of themselves in the process. Well, mistakes can happen, but you don't want to be the person yelling into the ear of an international colleague. You want to write in a professional, simplified tone, just like any other situation—the only catch is that you do so with the understanding that your reader sometimes speaks another language. An executive for a multinational computer corporation communicates via e-mail with Japanese executives on a daily basis:

"About 90% of my business communication is done through e-mail—I very rarely get to meet with my international colleagues face-to-face. So, e-mail *is* my face. And there is an obvious language barrier, which can lead to misunderstandings in the meaning of certain words or phrases. And I have no universal glossary of international business terms to reference, nor is there ever a bilingual colleague nearby when I need one.

So, I keep it simple and easy. Let's just say I don't use the term 'acetylsalicylic acid' when I mean 'aspirin.' I've also done a lot of reading up on the culture, and I try to mimic their salutations and closings so I don't offend anyone."

Tips for Writing to International Audiences

- *Keep the letter short and sweet:* Bear in mind the fact that your international reader may have to translate your entire message into his or her native language. By keeping the length to a minimum, and using short, basic vocabulary words, the translation is more likely to be understood and accurate—not to mention less time-consuming!
- *Be careful with physical dimensions and measurements:* If your message includes physical dimensions and measurements, understand that every country in the world, except the United States, uses the metric system. So, if you mentioned that the outside temperature was 105° today, you would mean *Fahrenheit,* but your reader might be concerned about ever visiting your location, since water boils at 100° Celsius in the metric world. To solve this problem, use both English and metric units in your message.
- *Be careful with dates:* Formats for dates and times vary greatly throughout the world. For example, in the United States, we would write the date, July 4, 2002 as 7/4/02. However, in Europe, this could be interpreted as April 7, 2002. And in Asia, it would be understood as April 2, 2007. So, be sure to read up on correct date usage for each new country that you e-mail, or ask them what form they prefer to use.
- *Be careful with time of day:* Time of day is another easily misunderstood item in foreign correspondence, and the global time differences that many people

work with every day don't make things any easier. The format most of us use for time (excluding the U.S. military) in the United States is not necessarily the same everywhere else. For example, if we set a meeting time for 3:45 P.M., our counterparts in Europe may be confused, as they use a 24-hour clock, rather than using A.M. or P.M. Better to just bite the bullet, and use a 24-hour clock when writing to a foreign colleague—it really isn't too hard once you get used to it. Or, you can always ask them what form they prefer to use.

- *Avoid technical phrases, jargon, and acronyms:* Although your reader may be an expert in a shared field, remember that he or she understands things in the context of his or her language, not yours. And technical terms are real bears when it comes to translation—many may not even exist in your reader's language. So, take the time to carefully explain whatever you think might throw your reader off.

- *Be sensitive to the intricacies of a different culture:* Although there is no way to know for sure what will offend someone—especially if you don't know them very well, or perhaps have never even met them—it is possible to be careful with your words. That rings true in any situation, foreign or domestic. But, you can stay on the safe side of writing by never including anything political, religious, or race-related in your writing. These topics don't belong in any kind of professional document anyway, so there is no need to include what doesn't belong in the first place.

- *Err on the conservative side:* It is definitely better to come across as too conservative, rather than as a boorish lout. Using slang terms or making casual references to people who can so easily mistake what you say is just asking for trouble. Keep your writing simple, polished, elegant, and relatively formal in all international correspondence. You can save your more casual rhetoric for when you have developed a stronger professional relationship with your international colleague. You can also take some cues from him or her—if your colleague is more casual, you can adjust your style to match.

The most important lesson to take away from this section is to develop awareness for others. Listen, learn, and become a more polished, more esteemed writer.

► SALUTATIONS AND CLOSINGS

> There are four ways, and only four ways, in which we have contact with the world. We are evaluated and classified by. . . . what we do, how we look, what we say, and how we say it.
>
> —DALE CARNEGIE

Dale Carnegie was right on all counts—how we are evaluated in this world is directly related to how we present ourselves. And, how you open and close your business correspondence will have a direct effect on your reader's perception of you as a professional.

The manner in which you address someone in a letter is a sensitive and important issue to consider for both you and your reader. Whatever you do, always remember to consider the appropriate salutation and closing by first putting yourself in your reader's shoes. You want to make the perfect impression—not too familiar, yet not too distant; not too bold, yet not too meek. You also want to make your reader completely comfortable with you from the beginning. That is where the salutation comes into play!

And the closing, or complimentary closing, as it is often called, is the last couple of words your reader is left with after finishing your letter. So, you want to leave your reader with an appropriate, professional, and warm feeling.

■

WORDS OF WISDOM
An often forgotten fact about closings
is that you only capitalize the *first letter of*
the first word in the closing.

■

Here is a comprehensive list of appropriate salutations and closings that should help you with almost any situation:

FEDERAL GOVERNMENT

Addressee	Salutation	Closing
President of the United States	Dear Mr./Ms. President:	Very respectfully yours, Yours very truly,
Former President	Dear Mr./Ms. _____:	Sincerely yours,

Vice President of the United States	Dear Mr./Ms. Vice President:	Respectfully yours, Yours very truly,
Cabinet Officers	Dear Mr./Ms. Secretary:	Sincerely yours,
Chief Justice, U.S. Supreme Court	Dear Mr./Ms. Chief Justice: Dear Chief Justice _____:	Very truly yours Sincerely yours,
Associate Justice, U.S. Supreme Court	Dear Mr./Ms. Justice: Dear Justice _____:	Sincerely yours,
Retired Justice, U.S. Supreme Court	Dear Justice _____:	Sincerely yours,
Speaker, House of Representatives	Dear Mr./Ms. Speaker:	Sincerely yours,
U.S. Senator	Dear Senator _____:	Sincerely yours,
Former U.S. Senator	Dear Mr./Ms. _____:	Sincerely yours,
U.S. Representative or Congressman	Dear Mr./Ms. _____:	Sincerely yours,
Former Representative	Dear Mr./Ms. _____:	Sincerely yours,
The Attorney General	Dear Mr./Ms. Attorney General:	Sincerely yours,
Chief Justice, U.S. Supreme Court	Dear Mr./Ms. Chief Justice: Dear Chief Justice _____:	Very truly yours, Sincerely yours,
Associate Justice, U.S. Supreme Court	Dear Mr./Ms. Justice: Dear Justice _____:	Sincerely yours,

STATE GOVERNMENT

Addressee	Salutation	Closing
Governor	Dear Governor_____:	Sincerely yours,
Lieutenant Governor	Dear Mr./Ms. _____:	Sincerely yours,
State Senator	Dear Mr./Ms. _____:	Sincerely yours,
State Representative or Assemblyman	Dear Mr./Ms. _____:	Sincerely yours,

LOCAL GOVERNMENT

Addressee	Salutation	Closing
Mayor	Dear Mr./Ms. Mayor: Dear Mayor _____:	Sincerely yours,

District Attorney	Dear Mr./Ms. _____:	Sincerely yours,
Board of Commissioners	Dear Mr./Ms. _____:	Sincerely yours,
City Council	Dear Mr./Ms. _____:	Sincerely yours,

JUSTICE SYSTEM

Addressee	**Salutation**	**Closing**
Chief Justice, State Supreme Court	Dear Mr./Ms. Chief Justice:	Sincerely yours,
Associate Justice, State Supreme Court	Dear Justice _____:	Sincerely yours,
Judge, Superior or Municipal Court	Dear Judge _____:	Sincerely yours,
Court Clerk	Dear Mr./Ms. _____:	Sincerely yours,

DIPLOMATIC PERSONNEL

Addressee	**Salutation**	**Closing**
American Ambassador	Dear Mr./Ms. Ambassador:	Sincerely yours,
American Chargé d'Affaires	Dear Mr./Ms. _____:	Sincerely yours,
Foreign Prime Minister	Dear Mr./Ms. Prime Minister:	Sincerely yours,
Foreign President Foreign Premier	Dear Mr./Ms. President: Excellency: Dear Mr./Ms. Premier:	Sincerely yours, Sincerely yours,
Foreign Ambassador	Excellency: Dear Mr./Ms. Ambassador:	Sincerely yours,
Foreign Charge' d'Affaires	Dear Mr./Ms. _____:	Sincerely yours,
Secretary General, United Nations	Dear Mr./Ms. Secretary General: Dear Secretary General _____:	Very truly yours, Sincerely yours,
Under Secretary, United Nations	Dear Mr./Ms. Under Secretary:	Sincerely yours,
U.S. Representative, United Nations	Dear Mr./Ms. Ambassador:	Sincerely yours,

| Foreign Representative, United Nations | Dear Mr./Ms. Ambassador: Excellency: | Sincerely yours, |

COLLEGE AND UNIVERSITY OFFICIALS

Addressee	Salutation	Closing
President or Chancellor, College or University	Dear Mr./Ms. _____: Dear Dr. _____:	Sincerely yours, Sincerely yours,
Dean of College or School	Dear Dean _____: Dear Dr. _____:	Sincerely yours,
Professor or Instructor	Dear Mr./Ms. _____: Dear Dr. _____:	Sincerely yours,

RELIGIOUS OFFICIALS

Addressee	Salutation	Closing
The Pope, Roman Catholic Church	Your Holiness: Most Holy Father:	Respectfully yours,
Cardinal, Roman Catholic Church	Your Eminence: Dear Cardinal _____:	Respectfully yours,
Archbishop, Roman Catholic Church	Your Excellency: Dear Archbishop _____:	Respectfully yours,
Bishop, Roman Catholic Church	Your Excellency: Dear Bishop _____:	Respectfully yours,
Priest, Roman Catholic Church	Reverend Father: Dear Father _____:	Respectfully yours,
Archbishop, Protestant Church	Your Grace: Dear Archbishop _____:	Respectfully yours,
Bishop, Protestant Church	Right Reverend Sir: Dear Bishop _____:	Respectfully yours,
Priest, Protestant Church	Dear Mr./Ms. _____: Dear Dr. _____:	Respectfully yours,
Minister, Protestant Church	Dear Mr./Ms. _____: Dear Dr. _____:	Respectfully yours,
Rabbi, Jewish Faith	Dear Rabbi _____: Dear Dr. _____:	Respectfully yours,

BUSINESS CORRESPONDENCE

Addressee	Salutation	Closing
Unknown Gender	Dear _____:	Sincerely,
Unknown Name	Dear Director:	Sincerely,
Corporation consisting of women and men	Ladies and Gentlemen:	Sincerely,

► RESUMES AND COVER LETTERS

"Make it thy business to know thyself . . . "

—MIGUEL DE CERVANTES

Although he most certainly wasn't referring to resumes and cover letters, Cervantes had an excellent point. It is important to know yourself well to be successful in life. And, when you are putting together something like a resume, that is designed to sell you to a potential employer, you need to do your homework!

In studying yourself, you simply compile all of your best professional attributes, all of your work experience, and your educational background, and get it down on paper. Try to include every major point that stands out in your mind, but weed out any unnecessary details. You can elaborate in the actual interview.

■

WORDS OF WISDOM
A great way to keep tabs on your own
professional career is to keep a work journal.
Jot down important events as they happen,
and be sure to include full dates and
all relevant information.

■

KEEPING TRACK OF YOURSELF

A work journal can be just like a personal journal and can be handwritten or typed, whichever is easier for you. Many busy executives are surprised at how much they can forget over days, weeks, months, and years, and have said that they do a much more accurate and thorough job of updating their resumes when they have a work journal to reference. So, it is not only an excellent resource for keeping track of projects or events, but it also enables you to build an impeccable resume.

With a work journal, you can bullet specific projects that provide details and a colorful description of you and your work, rather than just citing a bland list of statistics and duties. It will also help you keep track of any awards, certifications, specialized classes, training, or professional associations. Everything counts—and your ability to reference these specifics also demonstrates your level of involvement and mental acuity. As one executive says:

> Nothing impresses me more than to sit across the table from someone who actually knows what they've done in their own career. You wouldn't believe how many people I've interviewed who seem to be on autopilot when they're in the interview seat. I like to see a real, live, freethinking individual who is able to give me examples I can relate to. That tells me that "the lights are on, and this person is home."

In essence, you are giving yourself the credit you deserve for all of your hard work. Remember that *you* are ultimately in charge of your career, and it is up to you to take note of your own successes! And although you don't want to come across as a braggadocio, you can—and *should*—find a way to put your best foot forward. There is no shame in that, especially when it comes to resumes.

Sample excerpt from a work journal:

Date: Oct. 31, 20XX
Time: 10:30 A.M.
Re: AMEX/NorStar Meeting
Meeting Attendees: Bob Thornblatt, Trish Franck, Erika Kieser, Doris Gravrock, Donna Ammerman, Laura Jorgensen, Steve Johns.

Developed monthly budget projections. Agreed by all that I would calculate monthly revenues and expenses. Report due no later than the third of each month. Thornblatt named me lead accountant for statewide budget issues for the next fiscal year.

BUILDING THE PERFECT RESUME

Your resume is supposed to serve as an *outline* of your professional career, so focus on the highlights. Remember that employers are being bombarded with resumes, so yours needs to stand out.

- **Do your homework:** After you have compiled all of your personal and professional information (don't forget to refer to your work journal for details), then you can focus on researching the company with whom you want the job. You should also research the job itself—check all of the basic requirements and skills needed. Make sure you have included in your resume all information relevant to the job you are applying for.

- **Start with your contact information:** Your name, address, telephone number, and e-mail address should be listed first. Put your name in a slightly larger font size so it stands out.

- **Summarize your qualifications:** This is the initial summary that your potential employer will draw immediate conclusions about whether or not you are the right person for the job. The summary consists of several carefully crafted, concise sentences about your qualifications. The sentences should stand separately, and should be done using bullets, or by simply writing them out and ending them with a period.

- **State your objective, and be specific:** Your objective statement is a single sentence that reflects your specific goals. It is a statement of purpose that gives the employer an idea about what you plan to do with your skills if given the position. This makes it another key place for the employer to figure out if this is a potential match. The objective statement tends to be generalized, but it should not be vague, as in: *I'm looking to expand my horizons.* This is better:

  > Objective: *To obtain an executive-level position in the financial consulting industry, where I can put my 30 years of management experience to work.*

  Whatever you do, make your objective statement relate to the position you are seeking.

- **Organize:** Make sure that your resume is organized and easy to read. Resumes come in all shapes and sizes, but the most common formatting practice is to organize chronologically, starting with the present, and moving backwards in time. This is done because it highlights current skills and shows a logical progression of events. It is also common to list work experience first, followed by education, and finally, a brief list of professional associations (if applicable).

- **Keep it simple:** This means exactly what you think it does. Your resume is a *summary* of your qualifications and experience—and, although you want it to be a complete summary, just remember that you are not trying to write a novel.

- **Be thorough and accurate with the three basic components:** One executive refers to "education, application, and extras" as the "big three" in a resume—that means education, work experience, and professional associations. They have got to be there, and they have to be accurate. Don't embellish and don't drone on; but *do* be thorough.

- **Omit hobbies unless your interests somehow relate to the job you want:** You may have won distinctions or medals in your hobby that show a potential employer qualities like dedication, focus, or follow-through. For example, you may have trained for and run a marathon—this accomplishment shows potential employers you set challenging goals for yourself and succeed. Just make sure that there is an obvious connection between the hobby/interest and the job.

- **Omit personal characteristics:** Anything personal about you, like height, weight, race, religion, political affiliation, age, or gender should be omitted. Again, these factors have no bearing on your qualifications (the only exception is for something like a flight attendant position for an airline, where FAA guidelines regulate certain limitations).

- **Don't discuss salary requirements:** Leave the discussion of salary for when you are sitting at the negotiating table during the interview process.

- **Don't get wacky with color, size or set-up:** The very last thing you want to do is frighten a potential employer away with bizarre antics in your resume—that can only lead the employer to one conclusion about you as an employee . . . . As one executive says about resume paper:

> "If someone sends me a purple resume, it will definitely get my attention. But it will not get them the job. Stick with either classic white, a shade of gray, or an oatmeal color. Go with the ever-effective simple, classic, professional style."

  The same rules apply to the size and layout of your resume. Just send your resume on a regular piece of 8½″ × 11″ paper, and in a regular-sized envelope. As for the layout, keep it standard—anything else will cause the reader to toss it from lack of time to decipher it.

- **Toss all humility aside—it is your time to shine:** Although you should stay within acceptable professional boundaries, you are encouraged to shine in your resume. It is, therefore, the one place where you need to drop all humility. It is your chance to show a prospective employer that you are equipped with the skills and talents necessary to handle the job, and to even take it to a higher level than expected. So, while you don't want to make any promises you can't deliver, you can feel free to wax rhapsodic—in an honest way—about your skills.

■

WORDS OF WISDOM
Think of your resume as your moment in the sun.
Let golden beams of light dance off every
qualification and accomplishment.

■

RESUMES

High-Level Management Position

The following resume is that of an executive vice president. It is two full pages long, which is acceptable—even expected—for a high level position resume:

JIM WHITTLESEY
4340 Moondance Road, Cardiff by the Sea
California 92007
619-555-8645

PROFESSIONAL PROFILE

MANAGEMENT EXECUTIVE with 30 years' experience in a wide variety of assignments and projects demanding total accountability. A practical, realistic leader with the ability to get things done. Background includes:

- Operations Director at multiple locations, managing staffs and implementing a broad range of programs and strategies.
- Planned short/long range goals of growth, profit, and employee development.
- Provided sales and management consulting services to multiple industries with emphasis on goal orientation and accomplishment.

PROFESSIONAL EXPERIENCE

SAFELITE GLASS CORPORATION(1992–present). Regional Vice President.

- Responsible for sales results in a western 12 state area—$40 million in sales—50 employees.
- Directed field sales operation (West Region) resulting in a 25% increase in performance during 1993.
- Reorganized and developed regional management staff and field sales associates to embrace and utilize the concepts of expectation level, buy-in and personal commitment to produce results and obtain goals.

WHITTLESEY MANAGEMENT SERVICES (1987–1992). San Diego, California. Owner-Consultant. Clients included:

- AUTOMATIC DATA PROCESSING (1989–1992). Automotive Claims Services Group. Provided consulting services to include:
  —National account sales management; pricing/contract negotiation;
  —Client needs assessment; application/procedure analysis;
  —Management report analysis; management presentations.

- CHUBB INSURANCE GROUP (1987). Managed "bad faith" litigation activity to include:
  —Attending settlement conferences; recommending settlement value
  —Recommending appropriate discovery and analyzing results
  —Resolving fee schedule disputes; attorney selection and liaison.

MITCHELL INTERNATIONAL (1987–1989). San Diego, California. Vice President, Mitchellmatix Division.

- Executive responsibility included assignments in sales/marketing, client services, field support, and industry relations.
- Directed nationwide corporate sales and field support activity resulting in 74% over-plan performance during 1988.
- Served as corporate industry relations executive with insurance clients, trade associates, and industry organizations (1988–1989).

ALLSTATE INSURANCE COMPANY (1958–1986). Regional/Zone Claim Manager.

- Performed management assignments for 22 years in Oklahoma, Texas, Georgia, Florida, Washington, and California.
- Directed Claim Department operations for western 11 states. Responsible for results of 32 claim offices and seven house counsel offices—employing 2,200 people.
- Thoroughly experienced in organizational structure, salary administration, personnel management, budget preparation and employee training and development.

PROFESSIONAL ASSOCIATIONS

- Pacific Claim Executive Association (1981–1986)
- Advisory Board, California Fraud Bureau (Department of Insurance 1982–1984)
- American Insurance Association (Committee on Automobile Physical Damage 1987–1989)

EDUCATION

Southern Methodist University, Dallas, Texas

- Bachelor of Arts (B.A.)–1954
- Bachelor of Laws (L.L.B.)–1958

Mid-level Management Position

This resume belongs to a hard-working, mid-level manager. He is looking to climb a step up the corporate ladder, and has laid out his goals and experience in a simple and direct manner:

David John Pfaff
102 Azure St.
Sunnyvale, CA 94086
408-555-4243

OBJECTIVE: To obtain a position that would allow me to use my skills in supply chain management, while applying leadership and communication abilities.

SUMMARY OF QUALIFICATIONS:
- Excellent communications skills
- Self-starter, team player
- Able to adapt and adjust to changes
- Strong computer skills

WORK EXPERIENCE:

Dec 00–Present *Manager, Inventory Control & Distribution*
Sony Computer Entertainment America
- Manage all phases of receiving, distribution, inventory integrity, stock balancing, and physical count audits for five public warehouses.
- Responsible for reconciling all phases of SCEA's supply chain.
- Supervise a staff of one.

Apr 97–Dec 00 *Inventory Control Analyst*
Sony Computer Entertainment America
- Forecasted, tracked, and expedited all hardware and software receipts.
- Identified and resolved issues involving inventory, product quality, and warehouse procedures. Managed inventory levels, stock balancing, and inventory reconciliation for five public warehouses.
- Resolved shortages, damaged products, cost discrepancies, and performed physical count audits.

Nov 95–Apr 97 *Inventory Control Supervisor*
Performance First
- Supervised all company purchases and overlooked daily inventory control transactions.
- Responsible for accounts payable and governmental bid administration.

(continued)

PROFESSIONAL AWARDS:
"Recognition of Excellence" Award for the Seihan & Logistics Team
SCEA/SCEE/SCEI, June 1999

COMPUTER SKILLS:
Access, Word and *Excel, Business Objects*, and *Oracle* database application

EDUCATION:
M.B.A., Santa Clara University, 1993
B.A. in Business Management, Stanford University, 1990

CHANGE OF CAREER RESUME

The following resume landed a former Air Force navigator a job as an analyst and trainer with a global engineering firm. He needed to emphasize the number of flying hours he had in order to qualify for the desired position—so, he listed that information first, using several clear, concise statements. He also needed to emphasize the specialized experience he had in the particular aircraft the company was interested in—so, he listed that information in the last section of his resume to ensure they would remember it. This "first and last" combination technique worked:

Captain Stephen A. Coppi
61st Airlift Squadron
38 Reservoir Heights
Little Rock, AR 72337
Home Phone: 501-555-2748
Work Phone: 501-555-3198
navigator5@coppie-mail.com

QUALIFICATIONS
Highly qualified and experienced Instructor Navigator. Expertise derived from eleven years and 2,500+ hours of active duty flying.

Subject matter expert on C-130 tactics and defensive systems. Electronic Combat Officer.

FCF qualified; completed two separate FCF missions in Kuala Lumpur Malaysia.

C-130E 1,117 hours, C-130H 739 hours, B-52 H 582 hours, B-52G 116 hours, Total flying hours 2,690.

WORK HISTORY

1992-1994   *Instructor Navigator, 325 BS Fairchild AFB, WA*
Combat ready B-52 Instructor Navigator. Provided instruction to unit navigators on weapon systems and tactics. Developed tactical threat scenarios for realistic flight training. Hand picked for major flag exercises; flawless mission execution.

1994-1997   *Navigator, 52 AS Moody AFB, GA*
Combat ready, formation and airdrop qualified. Electronic Combat Officer, instructor for new defensive avionics equipment, provided instruction to aircrews on the use and performance of new onboard systems. As Tactics Officer, created and taught verification training for the entire squadron. Increased aircrew awareness of aircraft combat capabilities. Deployed during Operation Southern Watch, Saudi Arabia.

1997-2000   *Instructor Navigator, 517 AS Elmendorf AFB, AK*
Combat ready, formation lead, night vision goggle and airdrop qualified Instructor Navigator. Chief of Navigator scheduling responsible for scheduling over 40 navigators; provided a smooth transition as the squadron grew to become the largest C-130 squadron in the Air Force. Assistant Flight Commander for Readiness, responsible for the mobility functions of the squadron; directly led to the first ever "Outstanding" in the PACAF Unit Compliance Inspection. Deployed on short notice to East Timor in support of Operation Stabalise. Created a cyclone evacuation plan from scratch for the entire Air Force contingent deployed to Darwin air base. The plan was successfully executed as the entire group was forced to evacuate from the path of a cyclone.

2000-Present *Instructor Navigator, 61st AS. Little Rock, AR*
Immediately designated Chief of Tactics. Outstanding flying performance during JRTC at Ft Polk, LA.

EDUCATION

1984-1989	BA History, University Of Nebraska at Omaha
1989-1990	Undergraduate Navigator training, Mather AFB, CA
1991-1992	Combat Crew Training School, Castle AFB, CA
1994	C-130 Navigator Training, Little Rock AFB, AR
1995	Fighter Electronic Combat Officer Course, Eglin AFB, FL
1997	Squadron Officer School, Maxwell AFB, AL
1998	C-130 Instructor School, Little Rock AFB, AR

COVER LETTERS

An addendum to Cervantes' insightful words might be that it is also important to make it "thy" business to know the *company* to which you are applying, and the *position* you are seeking. That is what a cover letter is for. A human resources director for a global software corporation says that she simply tosses out resumes and cover letters that don't have certain key words in them. So, although you shouldn't fabricate skills and qualifications that you do not have, you *should* study the prospective company enough to know exactly what they are looking for.

- **Get the reader's information right:** Whatever you do, spell the name of the company correctly (believe it or not, people actually misspell company names in cover letters). Your resume will get tossed in the trash can immediately if you misspell the name of the company—that shows that you really don't care. You should also take the time to learn the name of the person you are sending your resume to. For example, if you are sending your letter to the Director of Operations, then find out her name, address her directly in your cover letter, and be sure to spell her name correctly:

    Dear Ms. Kleckner, is a lot more effective and personal than . . .

    Dear Director of Operations,

    Taking the time to find out this kind of important information demonstrates your dedication to detail and meticulous work habits.
- **Tailor your cover letter to fit the company and the job:** Think about your audience . . . who is your prospective employer, and how can your skills benefit their corporation? Do you have the relevant background needed? A vice president for a major flooring company says that one of the worst things he sees is a cover letter that has obviously been mass prepared. Of course companies real-

ize that many people apply for more than just one job at a time. And, of course you can't just invent a whole new background for yourself just to fit a particular job. But, you do need to tailor your letter for each individual company.

- **Be clear about the position you hope to get:** Don't beat around the bush about the job you are looking for. Simply state, "Dear Ms. Dafoe, I understand you have an opening for Dean of Students."

- **Tell why you are the right choice for the job:** Your cover letter is an opportunity to let your personality shine through your words, and help you stand out from the competition. Add particular pieces of information about your experience that describe you as the perfect candidate for the job.

  Don't say: I am a great motivator and team player.

  Do say: I led a team of six engineers through a software revision that increased productivity by 40%, and decreased down time by 30%.

- **Be specific about how you will make a difference:** This not only proves that you have done your homework about the company and the position, but it is a chance to highlight your attributes. If, for example, you are applying for a position as a recruiter, you can say something like:

  You advertised your need for an experienced recruiter—as my resume shows, I am that person. I can bring to the table five years of successful recruiting, plus a 95% retention rate within two years of placement.

- **Close the deal:** Once you have stated your qualifications, it is time to suggest the next step. Instead of the usual platitudes (like, *I'm looking forward to meeting with you soon*), one human resources executive suggests getting more specific:

  I'm available for an interview every other Friday, any time of day. I'll call your office on Thursday to discuss a possible meeting time.

SAMPLE COVER LETTERS

High Level Management Cover Letter

5478 Campbell Avenue
Tucson, AZ 85713

November 14, 20XX

Mr. John Weber
5546 McGregor Way
Merrimack, NH 03054

Dear Mr. Weber,

I've heard from John Walker in Human Resources that the position of Vice President of Operations for Datatech Corporation will open next month. Please consider this as my application for that position.

The enclosed resume outlines my extensive experience in operations. My many years in business have prepared me to deal with the added responsibilities of senior management and I am eager to take on the challenges of a position such as Vice President for Operations. My current position with Safelite Glass Corporation as Regional Vice President has prepared me to deal with the issues of operations in a large organization. I am ready to step up and assume the challenges of overseeing all operations in a large corporation such as yours.

While my tenure at Safelite Glass Corporation has been successful and rewarding, I am looking to broaden my professional scope, and I sincerely hope you will consider me for the position of Vice President for Operations.

I look forward to meeting with you and your organization. I will call your office Tuesday to schedule a possible meeting. You can reach me at the phone number and address provided on the resume. Thank you for your time and consideration.

Sincerely yours,

Ragini Mallavarapu
Ragini Mallavarapu
Enclosure

509 Somerset Drive
Spokane, WA 99210

August 20, 20XX

Mr. Steven Mays
10089 Wild River Lane
Spokane, WA 99210

Dear Mr. Mays,

I understand that the position of Corporate Inventory Control and Distribution Officer will soon be open in your company. I wish to be considered as an applicant for that position. I have included a copy of my resume that summarizes my qualifications for this position.

Throughout my professional career I have continued to advance to positions of greater responsibility, while at the same time continuing my formal education. I believe my combination of experience and advanced study makes me an excellent candidate for Corporate Inventory Control and Distribution Officer, as I can bring both experience and education to the Sony team.

In my time at Eastco Computer Entertainment, I have had the opportunity to meet with several of your inventory managers at professional conferences and seminars. On every occasion, they had nothing but wonderful things to say about your company. I know I can put my skills to work for your team at Sony.

Please feel free to contact me at home or work at the numbers and addresses provided on my resume. I can interview any weekday after 4 P.M. I look forward to meeting with you.

Cordially yours,

Brianna Wells
Brianna Wells
Enclosure

125 Fifth Street
Newton, CT 06001

May 10, 20XX

Mr. Michael Johns
2378 Summit Drive
Folsom, CA 95630

Dear Mr. Johns,

This letter serves as my application for the Training Instructor position within the military operations support division of your company.

I have included a resume that outlines my twelve years of experience as an Instructor Navigator in the United States Air Force. I have an extensive background in flight training, including experience in three major weapon systems: B-52H, C-130E, and C-130H. I have been involved in every step of training new aviators.

My decision to leave active duty and pursue a civilian life was a difficult one for my family and me to make. But, after many years of moving and being separated for extended periods of time, we decided that now is the time to get established in one place and to choose a field that involves limited traveling. While this decision means getting out of the Air Force, I don't want to leave the field of aviation that I love so much.

I strongly believe that my extensive experience in teaching aviators is a perfect fit for your needs. I am ready to change careers and go full steam ahead as a member of your organization. Thank you for your time and consideration and I look forward to meeting you. The contact information listed on my resume is current. I will call your office next Monday to set up a possible interview.

Sincerely yours,

Ella Jenkins
Ella Jenkins

The first and last thing to keep in mind about your resume and cover letter is that these critical documents are the first glimpse into *you* that a company gets. So, make it a good glimpse. Think first, write well, be thorough, be concise, be clear, be honest, and don't be afraid to go for the gold!

▶ PAPER AND ENVELOPES

> A blank page presents endless opportunities.
>
> —NED WILLIAMS

Your choice of paper and envelopes is the first panoramic picture of *you* that your reader sees. Even before reading the message, your potential employer is drawing conclusions about both you and your company—conclusions that are based solely on the physical qualities of the paper and envelope. However, while first impressions are important, not every situation calls for the highest quality, most expensive stationery. So you have some decisions to make!

There are four factors that can help you select the right paper for the right message: size, weight, type, and color. And, all combined, these things are what make your letterhead unique.

SIZE

The most common size of paper is letter ($8\frac{1}{2}$ by 11 inches). This is the standard business size in the United States, although some people use other sizes such as Baronial ($5\frac{1}{2}$ by $8\frac{1}{2}$ inches), or Monarch ($7\frac{1}{2}$ by $10\frac{1}{2}$ inches) for personal correspondence.

The weight of the paper is determined by how much a ream of paper (500 sheets) weighs. Most of the less expensive paper weighs between 16 and 20 pounds, while the high end products range from 60 to 80 pounds. As a rule of thumb, the heavier the paper, the higher the quality.

TYPE

Closely related to the weight of paper is the type of paper. The more economical types are made of sulfites. This type of paper can be used for mass mailings, interoffice memos, and routine informal correspondence. High quality paper contains some amount of cotton—and the higher the cotton content (also known as the "rag" content), the higher the weight and quality of the paper. Known as bond paper, the cotton content makes this paper feel smoother and heavier in the hands of the reader. Another quality you can choose is the type of finish with the paper. Smooth, rippled, and textured are just a few of the common finishes you can choose.

COLOR

Finally, the color you choose to use with your paper is important. White is the most common color because of its wide availability and the contrast it provides with black print. Other popular colors include charcoal gray, ivory, light blue, and pale green. Avoid choosing bright, fluorescent colors as they cast an unprofessional image on your message.

The bottom line is to select the right paper for the right situation. Interoffice memos most likely will use a 20 pound sulfite paper, white in color and 8½ by 11 inches in size. Formal correspondence to customers, potential clients, and financial institutions require a white or ivory colored bond paper with at least 50% cotton content. As a rule, always use the same paper size, weight, type, and color for the second, third, and subsequent pages as you used on the first page of your correspondence.

ENVELOPES

Envelopes, just like paper, are categorized by size, weight, type, and color. Standard sizes include Number 9 (3⅞ by 8⅞ inches) and Number 10 (4⅛ by 9⅛ inches). However, Number 10 is the preferred size for most all regular business correspondence. There are many other customized sizes of envelopes, but they are usually for personal or informal stationary. Just as you should use the same weight, type, and color on all pages of a message, you should also be sure the envelope matches the paper in these three categories. Lastly, the type of print used on the message itself is the only type of print that should be used on the envelope.

Sample Letters

Few things are harder to put up with than the annoyance of a good example.

—MARK TWAIN

This entire section is dedicated to providing you with good writing examples. In it, you will find realistic, quality sample letters for some of the most common situations you might encounter in business—including everything from acknowledgment letters to welcome letters. And, now that you are armed with powerful writing skills and effective formatting methods, you are ready to write quality business letters!

There are some basic guidelines that apply to *all* business letters you write. First, you need to follow all of the rules discussed in "Basic Training"—be clear, be concise, be mindful of your audience, use a positive tone, and never forget your goal. If you are writing a collection letter, for example, remember that your goal is to collect a payment, not to harass or anger anyone. If you are writing a sales letter, and you want someone to buy into your product or idea, then write persuasively and enthusiastically. Be thorough—don't leave anything out that might help tip the scale in your direction. Or, if you are writing a report, then write persuasively, using supporting statistics and pertinent information to strengthen your point. Use whatever tools you think you need to achieve your goal.

Remember that, in business, the bottom line is profit. And you cannot enjoy long-term profit without professionalism and goodwill. So, in all that you write, be professional, be clear, be tactful, be persuasive, understand your audience, and above all, keep your eyes on your goal!

Now, it is time to "get down to business" . . .

▶ ACKNOWLEDGMENTS

An acknowledgment letter is a letter you write to acknowledge some type of correspondence you have received. Proposals, invitations, requests, and applications may come to you in the form of a letter, e-mail, or a phone call, but all require acknowledgment from you.

Acknowledgment letters can be short or long. They can be a four-line response to a phone conversation or a comprehensive letter acknowledging the details of a proposal. On the next page is a letter acknowledging the receipt of a proposal. Remember, the purpose of this letter is not to accept or reject the proposal. You are writing to simply acknowledge you have received it.

12543 Remington Mine Road
Phoenix, AZ 85044

May 23, 20XX

Mr. Gregory Whipple
107 Cochran Avenue
Long Beach, CA 90803

Dear Mr. Whipple:

Thank you for your proposal concerning the redesign of our production facility in Phoenix, Arizona.

As you know, our company is expanding its operations in the western United States so we have many construction projects underway. I have assigned my lead project manager, Erik Bond, to the task of reviewing all construction proposals. In the future, all correspondence should be directed to him (wk: 505-555-3600; cell: 505-555-9208).

Once the proposal is reviewed, we will inform you of our decision. In the meantime, feel free to contact Erik if you have any questions or concerns.

Thank you for the professional proposal you submitted, and thank you for your interest in our company. We'll be in touch.

Sincerely,

Cathy Dafoe
Cathy Dafoe
President, Southwest Electronics

▶ ADJUSTMENTS

From time to time, "the system" fails the customer or client, and someone is in need of an adjustment. Broken merchandise, faulty products, refunds, and credits are all situations that require an adjustment. In short, when something goes wrong, it is up to you—the company—to fix it.

One thing to keep in mind about adjustment letters is that they should not be confused with *request* letters. A customer or client who is *seeking* help writes request letters, whereas the company who is *offering* help writes adjustment letters.

The letter on the next page outlines how a company will replace a broken product that was delivered to a customer's home.

1245 Terrel Lane
Midlothian, TX 76065

June 15, 20XX

Mr. Thomas Bulb
4413 East Lansing Street
Tucson, AZ 85748

Dear Mr. Bulb:

Our number one priority is customer satisfaction. So, when you notified our company about the malfunction with your new computer, our customer service department immediately went into full swing.

We have received your old computer, and have shipped you a brand new replacement.

Since you purchased your computer six months ago, our company has stopped producing your older model. So, the computer we have sent you is an all-new system, with upgraded software. I am also sending you a service agreement entitling you to one year of free technical support to help you get your computer and files back up to speed.

I hope that your new computer meets all of your needs, and that the manner in which we have handled this situation meets your expectations. As our valued customer, you come first. I hope we have regained your trust in our company.

Please call with any further questions or needs.

Most sincerely yours,

Ed Luttermoser

Ed Luttermoser
Vice President, Customer Relations

▶ ANNOUNCEMENTS

Announcement letters are used to let everyone know about changes taking place in an organization. Some announcements deal with personnel issues such as promotions or retirements. Other announcements pertain to changes in product development, sales strategies, or corporate policy. While all announcements are informative in nature, there are some rules of thumb to guide you to success.

Announcements should be brief and to the point, giving just enough details to get the message across clearly. Giving too much information about sensitive issues, such as controversial policy changes, can negatively affect employee morale. It also gives the impression that you are trying to disguise the issue in a flock of words.

Announcements about policy changes or broad strategies should also be kept short, leaving the detailed procedures for a formal memo. It is also important to keep the announcement balanced if it contains bad news. Be up front about any challenges facing the company, and include the course of action being used to deal with it. Remember that people tend to take bad news better if it is direct and honest, and is followed by a plan to conquer it.

The announcement on the next page describes how a company will deal with lower profits by instituting an unpaid vacation during the holiday season.

TO: All Employees
FROM: Brad Vase, President
RE: Holiday Vacation
DATE: April 24, 20XX

As most of you know, the economic downturn has affected our profitability for the last three quarters. This is resulting in a decline in revenue compared to our numbers from last year. To slow the decline of our bottom line, I have decided to close all operations from December 25 to January 2. All employees will receive unpaid holiday vacation time during that period. We will resume all operations on January 3.

This decision allows our company to end the year in the black, something that is a long-term benefit to all of us. It also allows employees some precious time with their families, for whom we are all very grateful. This policy will affect all employees, including myself, and all upper management.

I thank you all for your continued hard work—please enjoy some much-needed rest and relaxation. Next year promises to be prosperous and full of work!

Enjoy a happy, healthy holiday with your loved ones—see you next year!

Brad

Brad

► APOLOGIES

Business is a human venture . . . and along with humanity comes error. Letters of apology come in all shapes and sizes, depending on the situation. Apologies have to be made for things such as missed meetings, poor customer service, or misunderstandings. All of these instances require an explanation and apology, so that future problems can be averted and business relations can prosper.

The key to a good apology letter is sincerity. Nothing else in your letter will be heard if you come across as insincere. It is also important to be able to distinguish between private apologies and public apologies. If you have a private and personal apology to make, then keep it private and personal. If you have a public apology to make—resulting from a misunderstanding of policy, for example—then go public. In many cases, if one person has misunderstood what you have written, there is a good chance that other people may have misunderstood as well. So, write a helpful letter that apologizes for any confusion, and state that you hope to clarify and remedy the situation with your new and improved version.

Remember there is no need to fall all over yourself in an apology. Just be sincere and upbeat about how to rectify the situation. You can offer an *explanation* as to why a mistake was made, but don't confuse that with an *excuse*. Take responsibility. Then, sympathize with your reader and offer a solution to the problem.

6235 Anthony Drive
Edmond, OK 73003

March 23, 20XX

Mrs. Vivian Kiln
813 South Washington Boulevard
Eureka, CA 95501

Dear Mrs. Kiln:

First, I want to thank you for bringing our attention to the poor service you received on July 4th. There is no reason that you should have had to wait one hour to have your meal served.

As you probably remember, that day was especially busy, as it was a holiday, and so many people book their Independence Day dinners with us. Add that to the fact that the server assigned to your table was in his first week of training. Both the manager on duty that night and the server have been notified of this problem, and have pledged excellent service from this moment forward.

Please accept my sincere apologies. I would like to invite you and your family back to our restaurant for a complimentary dinner. I have enclosed a gift certificate that should cover the cost of your meal and even leave room for dessert.

Thank you for your patronage and your understanding. We at the Crow's Nest hope to see you soon!

Sincerely,

Vivian Davis
Vivian Davis
Owner/Operator

Appointment letters are short, straightforward messages giving the details about upcoming events or meetings. The basic pieces of information include date, time, location, people to be in attendance, and the reason for the appointment—this translates as *who, what, where, when, and why.*

4126 North Kolb Boulevard
Tucson, AZ 85750

December 2, 20XX

Mr. Wayne Coventry
6455 East Nightfall Drive
Sonora, CA 95370

Dear Mr. Conventry:

I look forward to our meeting on March 3rd at 9:30 A.M. My office is located on the corner of Sunrise and Kolb Roads.

Remember to bring your W-2s, receipts for donations, and last year's tax returns.

Please feel free to call me at 212-555-6035 if you have any questions.

Regards,

Madeline Martin

Madeline Martin
Accountant

► APPRECIATION

As necessary as it sometimes is to let employees know when they need to make improvements in job performance, it is equally necessary to let them know when they have done well. This promotes positive morale both in and out of the office, and also encourages even better work next time. So, show your employees that you appreciate their efforts by writing a quick letter of appreciation when it is deserved. It will make their day, and you will be satisfied with the long-term results of positive feedback.

---

300 Summit Dr.
Minot AFB, ND 58707

June 24, 20XX

Melanie Rosenberg
3421 Euclid Ave.
San Diego, CA 92123

Dear Melanie:

Please pass along my appreciation to your entire office at Headquarters for their outstanding efforts as members of the detail that prepared the base for the change-of-command ceremony on 23 May 20XX. Their hard work made the cleanup effort very successful, and it was evidenced by the fact that we received numerous compliments from senior leaders on the excellent condition of the base.

The success of events like these depends on the combined effort of an entire base and on the hard work from airmen such as Isaac Washington. Please congratulate the troops for a job well done.

Merrill S. Stubing
Merrill S. Stubing, Lt Col, USAF
Commander

▶ COLLECTION

From time to time, businesses run into the problem of customers or other companies not paying their bills. Oversight, financial difficulties, disputed charges, or the perception of poor service are among some of the reasons why people fail to meet their financial obligations. This is where collection letters come in.

Collection letters are sent in a series (if needed), with the first letter being the least severe—almost like a reminder letter. The letters then gradually become sterner until the final letter. This final letter is the most severe, as it usually announces pending legal action or other serious consequences.

There are a few things to keep in mind about collection letters. First, always give your customer the chance to pay the bill—your bottom line is getting the payment. You can even try to negotiate a slower payment rate if the first few letters have been ignored—whatever will get you your money. Second, your tone should never be emotional. No matter what the situation is, this is business—so, don't take it personally. Finally, never imply or state outright that your reader is a criminal or a liar—these words could get you in big legal trouble yourself.

This first letter is an example of a collection letter sent to a customer early in the collection process. The tone is informal and friendly, and suggests that the customer may simply have overlooked the bill. Remember to include a statement allowing the customer to pay the debt.

1847 Montgomery Avenue
Cerritos, CA 90703

February 11, 20XX

Ms. Jayne Letterman
2441 South Broadway Avenue
Columbia, CA 95310

Dear Ms. Letterman:

We were glad to see you in our store last month and even happier when you purchased the pine bedroom set. However, our accounting department tells me they have not yet received your payment for $645.67. I'm sure this is an oversight, so please regard this letter as a friendly reminder.

You have been a valued customer with excellent credit for several years, so I know this matter will quickly be resolved. If you have already sent the payment, please disregard this letter, and we thank you again for your patronage.

Sincerely,

Olivia A. Kirschman
Olivia A. Kirschman
Manager, Pacific Coast Furniture

If no response

If a customer fails to respond to the early letters regarding a bill, use a different approach. Instead of the "friendly reminder" theme, the objective of the next letter is to get the customer to acknowledge the issue and begin some type of correspondence.

The following letter presents a logical discussion of the situation. It no longer has the friendly tone of the earlier letter, and it mentions the next possible steps to be taken if there is no resolution.

1847 Montgomery Avenue
Cerritos, CA 90703

April 5, 20XX

Ms. Jayne Letterman
2441 South Broadway Avenue
Columbia, CA 95310

Dear Ms. Letterman:

According to our records, you did not respond to the first two letters we sent you regarding your bill for $645.67. It has been over 60 days since you purchased your furniture and we still have not received any payment. If you are having difficulty paying the bills, I'm sure we can arrange some kind of plan to pay your bill over a period of time.

Please respond to this notice, Ms. Letterman, as further delays will force us to pursue legal action and possibly damage your otherwise excellent credit. We look forward to your payment.

Sincerely,

Olivia A. Kirschman

Olivia A. Kirschman
Manager, Pacific Coast Furniture

If still no response

Most customers respond to collection efforts, and they pay their bills before legal action is taken; however, there are some who do not. These cases must be handed over to attorneys or credit agencies for ultimate collection. By this point, the customer will have been given every opportunity to pay the bill and should be notified of any impending legal action. Remember to keep your letter objective, stick to the facts, and avoid emotion. And, by all means, be clear about your next course of action.

1847 Montgomery Avenue
Cerritos, CA 90703

June 30, 20XX

Ms. Jayne Letterman
2441 South Broadway Avenue
Columbia, CA 95310

Dear Ms. Letterman:

Your account has been handed over to the Acme Collection Agency for your outstanding debt of $645.67. If you do not pay your bill within the next ten working days, the damage to your credit record could be irreparable. Please respond to this final letter, as it is your last chance to straighten out this situation before legal action begins.

I sincerely hope to hear from you so we can resolve this matter immediately.

Sincerely,

Olivia A. Kirschman

Olivia A. Kirschman
Manager, Pacific Coast Furniture

► COMPLAINTS

Whatever side of a complaint letter you are on (sender or receiver), remember the golden rule: *Never write to vent anger; write to get results.* Thomas Jefferson wisely said, "Nothing gives one person so great an advantage over another as to remain always cool and unruffled under all circumstances." This couldn't be more true than when it comes to your tone in business letters. There is something to be said for someone who can demonstrate grace in the face of adversity or pressure. And the "grace card" is the first card you need to play when dealing with a negative issue.

If you are the complainee, and you are at fault, then apologize for any problems or inconveniences that your mistake may have caused. Be empathetic in your tone, and feel free to briefly discuss where the problem began. This may help to avoid any future mishaps, which tells your reader that you are taking full control of the issue. Then, move on to the solution!

If you are the complainer, be firm and direct, but also courteous. And, if that doesn't solve the problem, then go straight to "the top"—the boss—with your issue. If you have kept accurate and thorough records, you can usually count on help from the person in charge. Be sure to use specifics in your description of the issue, and include all pertinent information, such as account numbers, dates, order numbers, contact names, or copies of financial transactions.

In the complaint letter on the next page, a manager in one department wrote to the manager in another department of the same company.

2815 Adams Street *Return Address*
Bellevue, WA 98005

September 29, 20XX *Date*

Mr. Barry Houseman *Inside Address*
1246 West Starr Boulevard
Palmdale, CA 93551

Barry, *Salutation*

We have seen some very large inventory discrepancies at the SMD facilities over *Body*
the past two weeks. These include:

On January 23, Bremerton shipped 2,304 units of 62403 (DVD remote) instead
of 62407. We are still trying to determine the customer.

Bellingham found an extra 160 units of 63400 during a cycle count. It was found
to be 1,425 units SHORT of 62407 at the EOM physical.

Fresno was found to be 144 units OVER of 62407 during the EOM physical.

These errors are huge, and the timing couldn't be worse. Due to the theft of the
two rail containers in New Brunswick our TA3 inventory will be tight throughout
the busy season. Because of this we can't allow almost 1,500 units to go missing.

Please take a look into this. I will be calling you next Monday to discuss and resolve
this issue.

Sincerely, *Closing*

Addison Brown
Addison Brown
Manager, Inventory Control & Distribution

► CONFIRMATIONS

Confirmation letters confirm deals, arrangements, or plans that have already been made. They are done to firm up an agreement in writing and also to allow for any changes or amendments.

There are five basic characteristics of a quality confirmation letter: 1) Be clear about what you are confirming—get specific with details; 2) Be personal—there is a good chance that you have an established business relationship with the person to whom you are writing; 3) Organize it well—be sure to list all information logically so it is easy to follow; this gives the reader a glimpse into how you will be handling all of your business dealings; 4) Don't go overboard—there is no need to rewrite a contract that has already been written; so, write only the highlights as reminders; 5) Use your good business manners—thanking customers for their patronage is a great way of saying that you will continue to do your best to keep their business.

3542 Wyndham Way
Newport News, VA 23602

August 12, 20XX

Mrs. Hilde Burnside
8776 W. Speedway Blvd.
Hampton, VA 23669

Dear Hilde:

Thank you for your continuing business. As discussed, Up & Down Elevator Company has agreed to the following in our contract for maintenance with the University:

1) The five-year maintenance contract, which was set to expire on March 31, 20XX, will be cancelled, and a new one-year agreement will be put into effect on October 31, 20XX.

2) The new agreement will have a one-year renewal clause, instead of a five-year clause. This change will not bind the University to a long-term commitment should circumstances change.

---

3) The annual maintenance charge will decrease from $16,700.00 to $10,440.00, effective October 31, 20XX.

4) The new payment schedule should go into effect as of October 31, 20XX.

Let me know what you think, and feel free to call if you need any further information on this new contract.

Best regards,

Patty Napaland
Patty Napaland

► CREDIT

Credit is a way of life in today's business world. Both companies and customers rely on a system of credit to purchase goods and services. Credit is founded on the idea that companies will take the risk of lending the customer money to make a purchase based on the understanding that the customer will pay them back with interest. The system is beneficial to both parties, but it is not without risk.

There are three kinds of credit letters: a request for credit, an acceptance of credit, and a denial of credit.

A customer who wishes to purchase goods from a company writes a request for credit. In business-to-business dealings, a "customer" who applies for credit is an entire company, and this credit application is referred to as *commercial credit*. For commercial credit requests, you need to explain exactly why you need the credit—if it is for office furniture, for example, state the make and model of the desired merchandise, the total amount of money needed, and the time frame in which you plan to pay off the debt. Be sure to also include references from other companies who have supplied your company with credit in the past. Be as specific as possible, including contact names and numbers of anyone who has worked successfully with your company regarding credit agreements.

REQUEST FOR CREDIT

5478 North Campbell Boulevard
Tucson, AZ 85713

November 14, 20XX

Mr. Ed Bricker
5546 North McGregor Way
Merrimack, NH 03054

Dear Mr. Bricker:

Our real estate company is refurnishing our office, and we are interested in some of the models we saw in your catalog.

We would like to order the following:

- 3 Cambridge desks (in dark cherry), each with a computer hutch
- 3 Executive desk chairs (in dark brown leather)
- 1 O'Brien printer stand

According to my calculations, these items total $2,650.00. We would like to purchase the above items on a line of credit with your company, for a 90-day term.

The companies listed below can speak to our financial standing. We have been doing business with them for over twelve years, and remain in good standing with all of them:

- Boynton Motors, 2615 East Broadway, Tucson, AZ 85712; 520-555-3312; Mr. Ted Cleaver.
- Sabino Information Solutions, 5565 East Sunrise Drive, Tucson, AZ 85750; 520-555-7777; Mrs. Lorraine Newton.
- First Bank of Arizona, 1522 North Kolb Road, Tucson, AZ 85711; 520-555-8888; Mr. Phil Collins.

Please let me know if you need any other information or additional references. I look forward to doing business with you.

Sincerely,

Marguerite Thatcher
Marguerite Thatcher

CREDIT ACCEPTANCE

It is always a pleasure to give someone good news—and that is exactly what a credit acceptance letter is for.

4156 George Washington Highway
Yorktown, VA 23693

May 4, 20XX

Ms. Elizabeth Thatcher
2164 Hunting Path
Yorktown, VA 23693

Dear Ms. Thatcher:

I am happy to tell you that we have accepted you request for credit. Your balance of $2,650.00 will be billed over a 90-day term. We can have your furniture delivered and assembled at your earliest convenience. Please call our main office on Bear Canyon Road to schedule a delivery date and time.

Thank you for your business! Be sure to call us for any future furniture needs.

Sincerely,

Adam Bricker

Adam Bricker
Credit Officer

CREDIT DENIAL

Unfortunately, not everyone who applies for credit gets accepted. Low income, outstanding debt, and past payment problems are common reasons why companies must occasionally say no. However, there are some things to consider when you write a denial letter.

The first thing you should do is thank the applicant for his or her interest in your company. Next, you must explain why you decided to decline the application. Be honest, but not brutal in your tone. Concentrate on financial status only, rather than on individual circumstances—remember this is business; it is not personal. Finally, let the reader know that he is a welcome customer on a cash basis, and when his finances improve, he is welcome to reapply.

---

901 Elmwood Street
Norwood, MA 02062

January 4, 20XX

Mr. Wilber Yates
1786 Blink Drive
Norwood, MA 02062

Dear Mr. Yates:

Thank you for your credit application and your interest in our products.

After much deliberation, we have decided that we cannot extend an offer of credit to you at this time. The income statement you provided, as well as your monthly expenses, leave little margin for additional financial obligations for you.

If your current debt becomes significantly reduced, we can reconsider our decision. Please consider that as a valid future option.

In the meantime, we hope we can service your needs on a cash basis.

Sincerely,

Deanne Powell

Deanne Powell

Employees, of course, are what constitute a company—they are the crux of the business world, the people who run companies. And they need to be dealt with, for good and bad reasons, on a daily basis. Writing something positive to your employees is far easier than writing something negative. So, we will focus on how to write a letter of reprimand and a termination letter.

REPRIMAND LETTER (EMPLOYEE)

Before writing a letter of reprimand, a supervisor should have already spoken to the employee in person. In fact, many such "counseling" sessions take place before a formal letter is written. If the verbal session fails to work, it is time to write the letter.

In a letter of reprimand, be sure to base the letter on poor performance, not on personality. This not only lessens the chance that the employee will become defensive, but it keeps you, the supervisor, in good legal stead. Again, this is a job, not a personality contest. So, focus on *job performance.*

3574 Pecan Street
Stafford, VA 22554

October 18, 20XX

Mr. Bob Rendell
475 Canton Street
Arlington, VA 22209

Dear Bob:

You have worked for me for over two years, during most of which time I have been pleased with your performance in the office. However, we've discussed on a number of occasions the amount of time you spend *out* of the office.

Until recently, your absences were within acceptable company limits. But, I am looking at records that indicate you were absent from work without first obtaining permission on ten different occasions in the last three months alone. Although your position as an Account Executive does require some days out of the office on client visits, it also means that your sales numbers should correlate. They do not.

Please consider this letter a warning, Bob. I know that, with determination and the right outlook, you can pull those numbers up in no time. So, let's work together as a team to get you back on track. You are a force to be reckoned with when you put your mind to it.

Come knock on my door anytime.

Hughes Lewis
Hughes Lewis

TERMINATION LETTER (EMPLOYEE)

Employees usually respond to well-written letters of reprimand. But, there are times when nothing seems to work for certain employees, and they need to be let go from the company.

Actual terminations are customarily done in person, but official letters documenting such a situation are necessary. As in letters of reprimand, you should remain focused on employee job performance.

1436 Thornlake Drive
Newport News, VA 23602

July 18, 20XX

Mr. Bob Huston
12236 Wilson Road
Williamsburg, VA 28188

Dear Bob:

This letter serves as formal notification of your termination from Pilgrim Coffee Company. We have held numerous meetings to discuss, and try to remedy, your steadily declining sales numbers and your failure to show up for work on a regular basis, but the situation has not improved.

After repeated counseling sessions and attempts to assist you in improving your job performance, I have decided that we will both be better served if you move on to another job opportunity.

I am truly sorry things did not work out, and I wish you the best of luck with your future plans.

Sincerely,

Hughes Lewis

Hughes Lewis

▶ FOLLOW-UP

A follow-up letter is common courtesy in business. It refers to a meeting or a phone call that has already taken place. It is a chance for you to offer further assistance or just reiterate a discussion or the terms of an agreement.

It is also a great way to "touch base" and maintain strong communication between you and your reader. It demonstrates that you are a thorough and efficient professional who will see a deal through to the end.

3421 Argonne Road
Papillion, NE 68046

December 3, 20XX

Mr. Chet Taylor
11445 Hall Boulevard
Las Vegas, NV 89128

Dear Chet:

It was a pleasure chatting with you this afternoon. Both Jackson and I were impressed with the breadth of your business experience and would be delighted to participate in your future.

As you requested, I'm enclosing literature about some of the products and services Highland Financial can provide you. This is not all-inclusive but should give you a good general idea of our capabilities. As I told you today, I believe that the primary consideration in any investment relationship comes down to the people working for you.

Chet, I'm delighted to represent Highland Financial, and I pride myself on my attention to my clients' goals and desires. I know Jackson feels the same way. I will be sitting down with a Portfolio Manager and Financial Planner to review our discussion. I will then call you to set up a meeting and discuss some of the options we can provide you.

Of course, I would be happy to answer any questions you might have in the meantime.

Best regards,

Nanda Navin

Nanda Navin
Vice President

▶ GOODWILL

Conventional wisdom says that you should "Praise in public and criticize in private." So when things are going well, say so. Letters of goodwill can range from congratulations on promotions to the civic involvement of your company. The audiences for these types of letters usually include members of your company, your customers, and/or the public at large.

Since these letters include good news, be upbeat but don't overdo it. Letters about someone excelling in their job should highlight their individual accomplishments but not alienate the rest of your organization. Announcing the volunteer work your employees have done can be a great public relations piece. The bottom line for all of these letters is sincerity. If you come across as self-serving or egotistical, your letter will do more harm than good. So, strike the balance between gloating and just plain pride in your people.

601 Somerset Drive
Spokane, WA 99210

August 20, 20XX

Ms. Hannah Wolfe
1008 Wild River Lane
Spokane, WA 99210

Dear Hannah:

Since joining our company six months ago, you have continued to prove yourself as a professional. Your ability to take on new tasks and get things right the first time make you a real asset to our team. I have talked with your peers, as well as other department heads and everyone has the same opinion of you: "an exceptional employee with great potential."

Therefore, it is my great pleasure to tell you that you have been accepted into the company's Executive Training Program. This is truly an honor, considering the high quality of professionals in your peer group.

I wish you great success in the program and expect to see you continuing to excel within our company.

Congratulations,

Lydia Walters

Lydia Walters

PUBLIC RELATIONS AS GOODWILL

The letter below is a *public relations announcement*—a type of goodwill letter—describing the community involvement of the employees in a company. The writer focuses on what the people involved accomplished, while at the same time tying their deeds to their company. Remember, be sincere and don't grandstand.

Technitron Employees Clean Up the Streets

Over the weekend, 20 Technitron employees spent several hours volunteering as part of OPERATION CLEAN UP. This project was started to help clean up the area around Harrison Park on the east side of Tucson. For the past six months, members of the locally based Technitron have spent one weekend a month picking up debris, fixing playground equipment, and renovating the original clubhouse built in 1952.

John McDonnell, a Technitron engineer, says that all the work is starting to pay off. "You can really notice an improvement in the park from the moment you drive into the parking lot. Since we started working here, more and more families have come to the park, not only to enjoy the sports fields and children's areas but also to help work along side all of the other volunteers." Rich Fields, Vice President of Operations at Technitron, says that giving back to the community is part of his company's commitment to Tucson. "We feel that being involved in our community gives us a connection to the area and allows our employees the opportunity to help solve problems right here at home."

If you have any questions about OPERATION CLEAN UP, or wish to help clean up Harrison Park, please call Brian Rogers, the Technitron Public Affairs representative, at 555-6666.

► GUEST SPEAKER

One day, you may find yourself in the position of choosing a guest speaker for an event at work. Here are some thoughts to consider when inviting someone to speak at your function: First, if your speaker is a known, public figure, then you will need to determine what that fee is, and whether your company can afford it, before inviting the guest to speak. You may also be in a situation where your company has little or no money to pay anyone to speak. In that case, be up front about your finances with the potential speaker and let him or her know exactly what you can and cannot pay for. If you can afford to pay for traveling expenses, then say so; if you cannot, let the speaker know in advance.

The most important thing to remember when inviting a guest speaker is to be very thorough in figuring costs and to maintain an open and direct line of communication with the speaker throughout the process. This way, there are no hidden costs or unfortunate surprises.

---

401 Surfside Way
Saratoga, CA 95070

March 4, 20XX

Dr. Harold No
2133 Loop Road
Hudson, NH 03051

Dear Dr. No:

The National Beachcombers Society will be holding its annual convention in San Diego, California from April 4–8, 20XX. On the evening of the 6th, the convention will host the annual farewell banquet at the Ocean Bay Resort, beginning at 6:00 P.M.

It would be our honor if you would accept our invitation to speak as our guest for the evening. Our program allows for a 30-minute segment that would be dedicated solely to you. The subject of the speech, of course, we would leave to your discretion. We will happily take care of your accommodations and travel expenses, and we also have a $500 honorarium to offer you.

We hope you will accept our invitation, as your attendance would make this evening a complete success! We would appreciate an answer by February 12th so we can make all necessary arrangements.

Thank you for your consideration. I look forward to hearing from you in the near future.

Very sincerely yours,

Anne Lennox

Anne Lennox

▶ HAVEN'T HEARD FROM YOU IN A WHILE

In order to maintain any successful business relationship, you need to keep in contact with people. Whether or not you think you will be doing future business with someone, it is important just to touch base now and then. You may find yourself with a hundred referrals from someone with whom you just kept cordial contact.

In a "Haven't Heard From You in A While" letter, you remind someone that you are still around and doing business and that they remain on your list of valuable contacts. You never know what may come up in the future, and maintaining this relationship can prevent awkward situations down the road where you will wish you had kept better track.

55 5th Street
Newton, CT 06001

May 10, 20XX

Mr. Andrew Mellon
2014 Summit Drive
Folsom, CA 95630

Andrew,

It is been a little while since the last time you and I were in contact with each other and you helped me find a solution for Walt Kim in the Volunteer Services office.

In any event, Steve Coppi spoke with my boss, Steve Johns, earlier today. Steve suggested that we contact you regarding your Internet and/or Intranet projects. As you may know, we've been working on Internet initiatives with Corporate Communications and several other entities within your organization for a number of years in a row. If possible, we would like to catch up with you on Internet/Intranet plans for the future.

When you have a chance, please e-mail me or call me (860-555-7777 ext. 11) to let me know when might be the best time for us to talk.

Thank you for your time, and I look forward to hearing from you.

Olivia Kirschman

Olivia Kirschman
Vice President, Internet Business Development

▶ INQUIRIES

When you are starting a project, or if you come across an obstacle in a project already underway, you may need to ask for help. It may be a simple request for information or additional resources, or it could be a complicated matter where you need to hire a specialized consultant to complete the job. Whatever your needs, you must begin by asking for help.

When writing an inquiry letter, keep a few basic things in mind: First, be as specific as you can when stating what you want—do so in the first paragraph. Put yourself in your reader's shoes, and remember that there is nothing worse than reading about everything *but* the question at hand. So, get to the point, use tact, and be specific.

Once you have established your needs, it is time to introduce yourself, especially if the person doesn't know you or your company. Give just enough background to help your reader understand what you need, but don't bury the reader with trivial details. Finally, since you can "catch more bees with honey," be sure to thank your reader for any help, and let him or her know how much you would appreciate any assistance.

5474 Sanford Drive
Phoenix, AZ 85302

September 2, 20XX

Mrs. Terri Green
6352 Douglas Street
Bellevue, WA 98005

Dear Ms. Green:

My company is expanding its operations in the Seattle area, and Mr. Tom Humphreys, our local real estate agent in Denver, referred us to you. We are looking for commercial property to lease on a long-term basis. We require at least 15,000 square feet of office space to house our new customer service center. And, we are looking specifically in the Bellevue area of greater Seattle. Our budget allows us to spend up to $8,000 per month. We are planning to move into our new offices no later than June 1, 20XX.

Our company is an online travel services firm. We've been in business since 1994 and are currently located in Phoenix, Denver, Atlanta, and New York. Our expansion into the Northwest has been in the plans for over two years and we are excited and ready to make the move.

Your assistance in finding commercial property will enable us to focus on the thousand other issues of opening a new office in an unfamiliar city. Thank you for your help—we look forward to hearing from you. You may contact us at the address or phone number at the top of the letterhead.

Sincerely yours,

Jim Lovell

Jim Lovell
COO, Welcor, Inc.

Choosing the right words is always important when writing letters. But it is even more important to be clear and specific when writing instructions. Regardless of what the instructions are for—applying for a loan, installing software, or operating machinery, for example—they must be absolutely understood by the reader. Leave nothing to chance. Don't use words that can be misinterpreted. Keep the language simple and straightforward. Even if adults will use the instructions, they should be written so a grade school child can understand them.

A key to writing instructions is to test them. In other words, write the instructions, and then try to decipher them yourself. Any ambiguous language or steps out of sequence should be corrected. Finally, give your instructions to a friend or a coworker to test, and make any necessary changes.

To: All Employees
From: Information Systems
Re: New Software
Date: November 28, 20XX

On December 12, the Information Systems (IS) department will install new software on the LAN. This installation will require you to reset certain parameters on your individual computer. Once you have logged on to your machine, please follow these instructions:

1. Go to Start menu and select Settings—Control Panel.
2. In the Control Panel window, double click on System.
3. In the System Properties window, click on the Hardware Profiles tab.
4. In the Hardware Profiles tab, click on the Rename button.
5. Type "New Configuration" and select "OK."
6. In the System Properties window, select "OK."
7. Close the Control Panel window.

When you have completed the steps above, log off your computer, and then log back on. All of the new settings should be in place.

If you have any problems or questions, contact Eleanor Staede in IS at extension 141.

Letters of introduction are written to introduce new people, services, or products. If you are introducing yourself, start with your name and your position in your company. If you are writing to existing clients, let them know you have reviewed their information, and wish to meet with them to discuss their business needs. If you are writing to a potential client, explain to them the products or services your company can offer, and again, ask to set up a face-to-face meeting.

When a company starts marketing a new product or service, it is usually a pretty big deal. All media forms are used: radio, television, newspaper ads, and letters. While some media reach a large audience (such as television), letters tend to be more focused, as they are sent to a more specific group of people. It is this focused approach that makes letters introducing products and services so effective.

1215 Fourth Avenue SE
Washington, DC 20330

October 24, 20XX

Mr. Jim Archibald
11489 South Craycroft Avenue
Washington, DC 20330

Dear Mr. Archibald:

My name is Allene Watt and I am writing to introduce myself as your new Investment Specialist with WorldBank Investment Services, Inc. I've set aside extra time in the coming weeks to meet with all of my new clients, and look forward to meeting with you as soon as your schedule allows.

I've taken the opportunity to look over your portfolio and I can see that you have a keen sense of strategic savvy. I also noticed that you subscribe to our suggested methods of diversified investing, which has obviously served you well. You can be very proud of your investment planning!

I would very much like to sit down with you to discuss the details of your future investment plans together. It will be a chance for us to get to know one another in person, and also a great opportunity for you to hear about some of the new stocks, bonds, tax-free investments, and other services we are now offering.

It will be my pleasure working with you and helping you to make sure your money continues to work hard toward all of your financial objectives. Please don't hesitate to call.

I hope to hear from you soon!

Sincerely,

Allene Watt

Allene Watt
Investment Specialist

▶ INVITATIONS

Companies use social events to conduct business in a wide variety of settings. Dinner parties at a country club, cocktails at a reception hall, or an awards banquet at a hotel ballroom are all possible scenarios for social/business events.

There are two types of invitations: formal and informal. Formal invitations are usually printed on heavy card stock, they are center justified, and include the term "R.S.V.P." (Répondez S'il Vous Plaît).

Informal invitations, on the other hand, are sent on company letterhead, with a name point of contact listed at the bottom of the letter.

The invitation below is a formal invitation for a business reception celebrating the ten-year anniversary of a small computer firm.

Tidewater Computer, Inc.
wishes to invite you to a reception
hosted at the Kiln Creek Country Club
Thursday, February Fourteenth
Cocktail Hour begins at 6 P.M., Dinner at 7 P.M.
R.S.V.P. James Bond, 555-1886

INFORMAL INVITATION

An informal invitation includes all of the same information about an event, such as place, date, and time. However, since it is being sent to people more familiar with the sender, the tone of the message is much less official than a formal invitation.

1215 Placido Domingo
Coronado, CA 92123

February 1, 20XX

Mr. And Mrs. Bruce Williams
24 Eastway Place
Coronado, CA 92123

Dear Bruce and Patty,

Caryn and I would like to invite you to a Spring Celebration that our company is hosting on March 23rd, at 5 P.M. aboard the Commander's Yacht. This is a casual party meant to open the spring sailing season with good cheer.

Social hour begins at 5 P.M., and dinner will be served at 6P.M. The group will be relatively small, comprised mostly of those of us who are avid sailors.

We hope that you will be able to make it, as it promises to be a wonderful evening. Please give our best to Matt, Chet, and their families.

Fondly,

Ben Ammerman

Ben Ammerman
619-555-2323

▶ MOTIVATIONAL

Everybody needs a little inspiration now and then—and people in the business world are no exception. With the pace of today's workforce, and the rate at which things change, sometimes a healthy dose of motivation is exactly what is needed to cure the ailing psyche. The letter below was written to a group of anxious employees as their company went through a large merger deal. Throughout the letter, their vice president of operations tried to encourage them to embrace change, not reject it. His goal was to inspire them to handle the situation with tenacity, fervor, and poise.

To: All Employees
From: Neil Barret, Vice President of Operations
Re: Change
Date: April 10, 20XX

The only constant in life is <u>change</u>. Change is real—change is today—it is tomorrow—it is your future. The consequences of change in your future hinge upon the quality of the choices you make today.

Change represents the <u>unknown</u>. Change produces anticipation, activity, anxiety, even annoyance, and avoidance. Change also produces a heightening of tension. But, don't fear change—address it, manage it, and learn how to make it work *for* you.

Managing change is <u>reality</u>—it is what we are required to do, both on and off the job. It is the manner in which we elect to manage it that is up to us. Making effective and positive choices in managing change is our ultimate challenge.

What if you <u>don't</u> change? You will stagnate and eventually regress. If you overly defend an aging status quo and prefer to live in the past, the world will pass you by. After all, the only people who want to go back to "those thrilling days of yesteryear" are Lone Ranger fans.

What if you <u>fight</u> change? Try to remember that constructive criticism is *good*—it is something that encourages you and allows you to grow and learn. Discussion and negotiation are essential—they are the only paths to a successful life. So, make the decision to embrace change, not sabotage it. Commit to it.

<u>Changes and choices</u>—they are inevitable. We cannot escape them. We must make the necessary choices to not only cope with them, but to master them. It is up to you. What choices will you make?

► ORDERS

When you purchase something from a catalog or though the mail, the seller will often have some kind of order form for you to fill out and mail in. With advanced technology, people now often e-mail the same information. Sometimes, however, there may not be an order form to use; so, in place of an order form, you can send a letter.

Order letters must be very specific. Start with the item you want to purchase: Does it have a model number? Does it have a unique name? Be exact—include the price of the item if it is known. Specify the quantity you want as well. Explain what method you will use to pay for the goods—will you be using a check, credit card, or line of credit? And, of course, include where you want the goods to be shipped. If the billing and delivery addresses are not the same, be sure to specify which is which.

2624 Elk Creek Road
Seattle, WA 99223

November 23, 20XX

ACME Supply Company
24563 West Industry Way
San Diego, CA 92121

To Whom It May Concern:

This letter serves as my order for the following merchandise:

1—Expersion Coffee Maker—Model # 144FG	$49.99 each
1—Italian Espresso Machine—Model # 267AH	$59.99 each
1—Coffee Bean Grinder—Model # 523EE	$25.99 each
	Total: $135.97

Please charge this entire order (including tax and shipping) to our corporate credit card: 1111-2222-3333-4444: Expiration: June 20XX.

Our shipping and billing addresses are the same:

HCO Enterprises
2624 Elk Creek Road
Seattle, WA 98195
509-555-1313

Thank you!

Caryn Fetzer

Caryn Fetzer
Human Resources Manager
HCO Enterprises

▶ PROPOSALS

A proposal is a persuasive letter geared toward selling your ideas to someone. It is not a mere presentation of information, but it is a call to have your ideas put into action. It is typically lengthier than the average business letter, and should include whatever information is necessary to substantiate your needs.

A proposal can be used for a number of different reasons: you may need funding for a project; you may want to propose an idea to your company; or you may be seeking a contract. Whatever your reasons are, there are a few key elements that need to be included in an effective proposal:

- Clearly state your intentions—define your idea so that there is no room for misunderstanding.
- Be specific when making a case for your idea, and when explaining how you will follow through with your plan.
- Offer solutions to any potential problems before they are brought up.
- Provide step-by-step directions as to how to proceed once the proposal is accepted.

You can choose whatever proposal format best fits your idea in size and scope—shorter proposals for smaller-scope ideas, and longer proposals for larger ideas. Use graphs to illustrate your points (if needed), or choose from among any of the formatting and writing tools offered throughout this book. And, just like any other piece of business correspondence, remember to be clear, thorough, and mindful of your reader.

CLIENT	Work & Company
JOB	Collateral Kit, Stationery
CONTACT	Judy Meagles
DATE	September 26, 20XX

OBJECTIVES

A. To create camera-ready artwork for Work & Company Inc. for their business cards and Canadian stationery. Design will match existing business cards and stationery with minor revisions to type and layout.

B. To create a collateral kit—envelope, one-color pocket folder, and two-color four-to-eight page brochure—for Work & Company Inc., which will match the existing look of the stationery system. Work & Company will supply George Brown Design (GBD) with text on disk for use in the brochure. GBD will, if necessary, work with Work & Company to streamline and focus the existing brochure copy. For the First Round Presentation, GBD will present two sample designs of the brochure and a mock-up of the folder. Second Round Presentation will be a revised version of the brochure chosen in round one. Folder mock-up will be similar to the existing report cover. GBD will negotiate print pricing and handle prepress and print production (including press-check) to ensure successful completion of Work & Company's jobs.

PRICING

Business Cards and Stationery

DESCRIPTION	HOURS	HOURLY RATE	TOTAL
Layout and Production	5	$55	$275
Total			$275

Folder, Brochure, and Envelope

DESCRIPTION	HOURS	HOURLY RATE	TOTAL
Design	16	$85	$1,360
Layout and Production	8	55	440
Administrative (print-pricing, etc.)	8	45	360
Text Revisions	hourly	40	TBD
Total			$2,160+

Schedule

First-round presentation	TBD
Second-round presentation	TBD
Final for proofing	TBD
Final to Printer	TBD

▶ REASSURANCE

A letter of reassurance is sometimes just what is needed to help ease the mind of concerned clients. It shows that you have their best interest at heart, and that you *care*. It also makes them feel comforted to know that they can rely on your skills and expertise to lead them through a difficult time. It is particularly important to reassure clients when an event affects many lives across the world.

Although letters of reassurance are used in any and all industries, the financial world is especially hard hit as our stock market reacts to the daily whims of our dynamic global economy. So, a carefully crafted letter of reassurance is both a comfort and an effective way to keep your clients positive and forward-looking.

1247 Charles Street
Escondido, CA 92026

February 22, 20XX

David and Vivian Blalock
236 East 42nd Avenue
Hampton, VA 23666

Dear David and Vivian,

There have been many economic ups and downs this year.

As we move into the future, I am optimistic. Throughout our hardships, we have proven how resilient we are. I believe that, in the coming years, we can move closer to profitability. It is my hope and my determination that we will enjoy renewed prosperity for years to come.

I continue to see opportunities ahead in 20XX. Changes in our economic strategies can allow businesses to expand, and new employment opportunities to crop up for those in need. In light of those changes, the stock market will be better equipped to react more positively. And, economic resurgence can be translated directly as portfolio growth.

While I cannot predict the future, I can assure you that I will continue to provide the best possible services and guidance throughout 20XX and beyond.

Please call me anytime to discuss your needs—I know we can work well together to make the most of your investments. Thank you for your continued support and friendship, and best wishes for a happy, healthy, and prosperous New Year!

Very truly yours,

Deborah Brizgy
Deborah Brizgy
Chief Investment Strategist

▶ Recommendations

A recommendation letter is written to endorse an idea, a project, or an opinion. It can be sent to just one person (a client, for example), or to a group of people (an organization). In a recommendation letter, you are essentially trying to sell your ideas, so you will use many of the same techniques that you use in a sales letter.

Use clear headings that lead the reader comfortably through the process. Use a few sentences to describe the need or the basis for your recommendation, and then state it outright—you don't want to leave the reader guessing about your point. Since most recommendations involve some degree of change and uncertainty, be sure to buffer the sting with a good description of the payoff of your plan.

884 Doolittle Lane
Yuma, AZ 85364

May 17, 20XX

Mr. and Mrs. Brown
2165 Dodd Boulevard
Tucson, AZ 85749

Dear Joe and Lori:

I am happy to report that Montes and Associates did better than can be expected
considering this past year of economic upheaval and uncertainty. And, we are
continuing to climb back upward! Because of this trend, I want to take a moment
to summarize the results of the past 12 months and also make you aware of the
many financial options at your fingertips.

Fiscal Results

Montes and Associates finished first for the trailing 12-month period, third for
the trailing five years and tied for fifth for the September quarter among ten major
firms in the quarterly stock-picking competition tabulated by Banks Investment
Research Inc. and reported in *The Financial Times*. Overall, Montes' strategy has
ranked first for the five-year period in 20 of the 51 quarters it has been in the
competition.

We are, of course, very pleased with our results during the September quarter. But,
we should remember that stock picking is not a science, nor an art. At best, it is a
blend of the two, along with a sprinkling of sheer luck. Even the most successful
analysts, with the most regimented strategies, will be wrong some of the time.

Investment Strategy

As you well know, the investment approach Montes and Associates prefers is diversification. We live by the guiding principle: *Don't put all your eggs in one basket.*
We recommend a balanced portfolio of stocks, with the proportions dictated pri-
marily by risk factors and scope of time for each client.

I like to remind my clients that investing is a long-term venture and that time
has a way of making most investments look good. But, the investment must first
have strong fundamentals and a sympathetic atmosphere.

Recommendations

Investing in individual stocks is not for everyone. It takes strong risk tolerance, a solid understanding of the cyclical market, and the ability to maintain perspective. For most investors, only a small portion of their portfolios should be invested in more provisional individual issues. A more conservative, professionally managed distribution for the rest of their portfolios should help balance both real and paper losses in individual issues.

Remember that past performance is not a guarantee of future results. And, of course, individual investor results will vary—achieving similar performance figures is not likely. As we have seen this quarter, many securities recommended by Montes and Associates are subject to wide price fluctuations. *The Financial Times* does not sponsor or endorse Montes and Associates' investment plans.

For more detailed information about our recommendations, or if you need to discuss your portfolio, please feel free to contact me.

Very truly yours,

Cade B. Kirschman

Cade B. Kirschman
Senior Securities Analyst

► REFERENCE

Reference letters are "additional information" letters that accompany a resume. Their purpose is to familiarize a potential employer with the more personal side of a candidate—something that bulleted facts on a resume cannot do. The focus should still be on those traits that compliment the professional work life of the individual, but anything exemplary is fair game in a reference letter.

If someone has asked you to write a letter of reference on his or her behalf, it is a great compliment. It also means the person is expecting that you will do them justice in your interpretation of their qualifications. So, ask them what they would like to emphasize, hit all of the highlights of their life and career, and write a letter that cannot be ignored.

3535 Nightengale
Crystal Lake, IL 60554

October 31, 20XX

Golden Star Management Services
1001 Sundance Circle
Phoenix, AZ 85044

To Whom it May Concern:

This letter is intended to serve as both a personal and professional recommendation for Dave Ammerman. You may have had an opportunity to review Dave's resume, so I won't rehash the details of his work history. Rather, I will furnish you with my personal observations of Dave, spanning the last 19 years.

I had the opportunity to work with Dave for ten years, during which time we developed a close business and personal relationship. Dave maintained a direct reporting relationship to me for seven years in a variety of management assignments. His attitude was always upbeat and positive; his work was unexcelled; his loyalty was unquestioned; and his performance exceeded expectations.

Dave possesses all the usual business qualifications in education, training, technological expertise, and experience. However, his attributes far surpass the usual. Dave is a quick study. He has the analytical ability to size up a situation and take whatever action is necessary. Dave readily accepts a challenge, possesses unbounded enthusiasm, and approaches everything he does with a good sense of humor.

Through his early years of development in a military family and his graduation from the U.S. Air Force Academy, Dave learned discipline, respect, and the importance of being team oriented. Through athletic participation, pilot training, and military combat experience, he refined his own personal dedication to achieve his full potential.

Dave's business philosophy can be best demonstrated by these three key traits:
Initiative—He is always exploring a better, more effective path.
Creativity—He thinks outside the box when developing solutions.
Flexibility—While remaining committed, he has the common sense to alter course/directions if the facts so dictate.

Dave Ammerman is a thoroughly proven business executive.

It is my sincere pleasure to recommend Dave Ammerman. If you require any further information, please feel free to call or write me.

Jim Whittlesey
Jim Whittlesey
3535 Nightengale
Crystal Lake, IL 60554
804-555-6500

A refusal letter is essentially a letter telling someone "no." In business, you might have to say no to things like granting a contract, hiring an employee, using certain services, or buying particular products. Above all, your goal is to say no gracefully and professionally.

For example, if you can see a potential future with the involved party, but you are truly limited at the time you are making a decision (your hands are tied), then write the letter with that thought in mind. Use an open-ended tone. However, if you honestly see no future—ever—then you can politely suggest alternatives, thus providing help to your reader. Your letter should be as positive as possible, with the actual refusal stated only once, and very briefly. The rest of your letter should be courteous and reader-oriented.

Many people believe that recipients of refusal letters also deserve some sort of explanation as to why they were refused. If you take the time to explain why you have made your decision, then you are serving two purposes. You are helping your reader come away with a better sense of what to do next time, and you are helping yourself avoid having to answer more questions down the road—no need for the situation to linger and brew. Keep in mind that your decision should be fact-based, not emotional. This is business, and your decisions should never be personal. And, of course, appeal to your reader's common sense and business knowledge—it will give him or her the feeling of being appreciated and understood.

So, remember: be tactful and positive, and above all, write your refusal letter with a "smile in your words." It is all in your delivery.

3421 Hall Boulevard
Tucson, AZ 85708

November 17, 20XX

Ms. N. Gumby
145 Dogwood Drive
Papillion, NE 68046

Dear Ms. Grumby:

We were pleased to receive your proposal for the "Fish-O-Matic" in response to our call for ideas for a revolutionary kitchen appliance.

Your idea is an interesting one; not many people would have the vision to design an appliance that transforms whole raw fish into "wholesome fish cakes." While

---

our engineers evaluated your design, they were particularly struck by the ingenuity of your "bone removal device," and we were very curious as to exactly how it would function given what we know of Newton's Laws of Motion and fluid dynamics.

Unfortunately, we are unable to accept your design proposal. Our company is seeking design ideas for more conventional appliances, and our current production lines are not tooled for such radically new ideas. When we put out our request for proposals, we were looking for improved designs for existing household appliances like can openers, dish washers, and the like. We are also concerned that the market share for such a product, outside of Louisiana, might be limited.

The "Fish-O-Matic" does seem to have potential, and we would suggest you solicit our competitor, GadgetCo, to see if they have any interest. We would encourage you to be vigorous and persistent in your proposals as they may be reluctant to take on visionaries such as yourself.

Sincerely,

Luke Christopher

Luke Christopher
Chief Engineer
Alliance Appliances, Ltd.

▶ REMINDERS

Reminder letters are used to remind employees about events, meetings, or special occasions. Since these things are usually set up well in advance, reminder letters help to solidify plans and to make sure all involved parties will be aware and/or be in attendance.

To: All Faculty Members
From: Campus Administration
Re: Fire Alarms
Date: May 15, 20XX

On Sunday, January 23rd, the campus fire department will be checking the fire alarms in all buildings. These tests will begin at 9 A.M., and should conclude by 5 P.M.

If you have any questions, contact Jim Ellis, Campus Fire Department, at 555-1212.

► REPORT

A report letter is written as a summary of a more in-depth study. A report letter summarizes in one page—or two pages maximum—what a three-plus-page actual report states. Reports describe the *outcome* of an operation or a study. They do *not* serve the same purpose as a business plan, which discusses future operations. Reports are very structured, usually include both how statistics and information were gathered and what your conclusions and recommendations are.

There are many different types of report letters in business, including reports on employee retention rates, the success rate of employee medical insurance programs, corporate improvement suggestions, departmental or company-wide changes, and progress and financial reports (the most common). Financial reports are so widely used that they all typically follow a standard set of rules. First, and most importantly, financial reports are about money—so, use numbers, rather than excessive text. Second, use whatever you need to give a visual of what you are asserting—this includes using charts, graphs, and tables. Next, be sure to use specific percentages rather than vague statements. It also helps to attach additional notes or information for further clarification—remember that this extra information should not be included in the actual letter. And, as always, use standard terminology—your goal is to garner total understanding by everyone, not stump people with unclear information.

On the next page is a report letter that summarizes findings and gives recommendations about a new product proposal.

1001 Sundance Circle
Phoenix, AZ 85044

January 15, 20XX

Mr. W. Christopher
3535 Nightengale Drive
Crystal Lake, IL 60554

Dear Mr. Christopher:

We recently completed the design testing for the new dishwasher line and this report summarizes our findings.

We evaluated the new prototype based on five main criteria: durability, ease of use, performance, energy efficiency, and manufacturing cost. The product scored well in all areas except manufacturing cost. We believe the new product needs some more engineering before it is ready for production. Our calculations show this dishwasher could be very costly to produce (at least $650 per unit). I'll address the findings in each criterion below; the full engineering report is attached.

Durability was a strength for this product. The design calls for a steel cage design for the construction and faceplate. By using light gauge steel all around, we can expect the product will look good and protect the machine parts throughout the service life (10 years). Interior machine parts are a combination of composites, steel, and high-density polyethylene (HDPE). These materials performed well in our wear tests, and we had no parts malfunctions or breakage during the service life test. We also feel the steel construction is very attractive and lends itself to some very durable and clean surfaces. **Grade A**

This product is very easy to use. The control system was very user friendly, kudos to the Human Factors Team in Product Design. Controls were well laid out and the icons were easy to understand. We also found the draft user manual helpful and well written. We suggest a subject index in the back for troubleshooting versus the current table, however. **Grade A**

Performance was exemplary. We tested against all types of food, including baked on tomato sauce, and the machine cleaned everything fully. We further tested with five different types of detergent with no marked difference in performance. **Grade A**

Energy efficiency was good, but there is room for improvement. The prototype used 3.4 kwh per wash, above average for an appliance of this size. Additional insulation in the cage and around the heating element would improve efficiency markedly. **Grade B**

Manufacturing cost is excessive. The same features that make the prototype durable also increase the cost. While the carbon fiber composite materials save shipping weight, we recommend using a light gauge steel or aluminum in their place to save cost. In the attached report we detail the specific material change proposals that we feel will lower the unit cost by $100. **Grade D**

We found this prototype to be one of the best New Product Design samples our company has seen in some time. If manufacturing costs can be reduced through materials substitution or some other means, this product should be a great new line.

Sincerely,

Elizabeth Haley

Elizabeth Haley
Chief, New Product Design Department
Alliance Appliances, Ltd.

► REQUESTS

Aside from sales letters, request letters are among the most commonly written letters in business. And, there is a wide range of topics that fall under the request umbrella. A request letter can be anything from a letter asking for a brochure or a product to a letter asking for information about a company. If you are the *sender* of a request letter, it is best to get to the point. Ask for what you want clearly and concisely, and thank your reader in advance for his or her time and consideration. If you are the *recipient* of a request letter, be sure first to thank your reader for his or her interest when you write your reply, regardless of whether or not you can complete his or her request. If you are responding with the information or materials requested, refer to the original request in your letter and list all accompanying documents or materials.

Requests from one business to another usually involve one employee asking another employee for something—like assistance with a project. The letter below illustrates one employee's request for the assistance of her counterpart in another company. So, she maintains goodwill and a strong working relationship by writing a succinct, positive note with bullets to help outline her requests.

---

420 Crabapple Drive
Yorktown, VA 23692

June 18, 20XX

Mr. Jeff Sutton
5243 South Portage Road
Newport News, VA 23606

Dear Jeff:

Please find enclosed materials for the Spring-Summer issue of AFS's newsletter: a Zip disk with all the files, photos, artwork, fonts, and a color dummy of the Newsletter and Scientific Article pullout.

A few notes:
- This issue has an additional spread, as it is a double issue, so it will total 16 pages instead of 12.
- Nothing needs to be scanned at this time; all of the photos and artwork are on disk, placed in the QuarkXPress file, and ready to be output to film.
- The Scientific Article pullout is front and back of one page.

- Steve Perry is no longer with AFS, so unless otherwise noted, Gordon Sumner will be your contact for quantity, delivery date, blues, and so on.

That's about it. I'm sending this to AFS prior to sending it to you. So, if there are any corrections, please input them prior to sending film. Call if you have any questions. Hope all is well.

Best regards,

Renee Worden
Renee Worden

Whether you are making travel arrangements or scheduling meetings, reservations are imperative. Limited seating and scarce availability are two issues that require advanced planning. Making reservations will give you peace of mind and allow you to concentrate on the critical details of your event.

---

455 Timberwood Lane
Shreveport, LA 71104

August 1, 20XX

Ms. Andrea Poppins
9663 Harris Drive
Spokane, WA 99223

Dear Ms. Poppins:

I would like to reserve the Tahoe Room on June 10th, from 8 A.M.–4 P.M. We will need seating for 30 guests, one podium with a microphone, a small table for a Proxima projector (we will provide the projector), and, of course, a large screen.

Please bill our account (#2376), and also call me at the number below, when you confirm our reservation. We have held many events at your hotel, and I am certain this one will be another smashing success!

Thank you,

Gilligan Smith
Gilligan Smith
520-555-1591

▶ SALES

It can be argued that almost every business letter is a sales letter in some respect—with the exception of letters that are strictly informative. In fact, in most business correspondence, you are either trying to sell something or someone (i.e. reference letters). Just by virtue of your *writing* a letter, you are extending a goodwill gesture, the ultimate purpose of which is to sell your good image to the reader.

One important thing to remember about sales letters is that they are one of the exceptions to the "keep it short" rule. If your reader is going to invest in something, large or small, he will need to know as much as possible in advance. So, this is a time where a longer letter is not only more appealing, but it is also more effective. This doesn't mean you should ramble—you still have to engage your reader and make every word count. But, in a sales letter, you can use as many words as are necessary to accomplish your goal.

Use the techniques that the pros use: catchy phrases that grab the reader's attention, plenty of descriptive details, lots of supporting facts (to establish need), and unbeatable offers. Keep in mind the well-known advertising principle, *AIDA: Attention, Interest, Desire, and Action*. In other words, get the reader's attention, retain his interest, tune into his needs, and call him to action.

3268 129th Avenue
Fair Oaks, CA 95628

October 5, 20XX

Mr. and Mrs. Cox
6635 Boxwood Drive
San Antonio, TX 78251

Dear Mr. & Mrs. Cox:

As your Nikula-Bentley Financial Advisor, I make every effort to be a valuable source of financial guidance for you and your family. Part of my mission is to help you remain committed to your long-term financial plan, even in the midst of uncertain economic conditions, like the ones we all now face. It is at times like these that I feel it is especially important to maintain a diversified portfolio.

The financial plan we have built for your family considers your personal situation, and includes specific types of investments specifically designed to suit your needs. However, I realize that as analysts talk of a possible economic recovery, moving toward individual stocks may be tempting—of course, we all hope to garner the benefits of potential upturns. But, for most investors, individual equity securities should be only a small portion of their broad portfolios.

So, it is especially important to make informed decisions about the stocks you choose. My input, along with the support of knowledgeable, trained analysts, and intensified research from Nikula-Bentley & Associates, may help you to feel more confident about your choices.

One way I can help you with stock-picking decisions is through our Financial Research Department's annual "Hot Picks" list. This list highlights a diversified portfolio of individual stocks that we expect will produce excellent results this year.

Our Financial Research Department is highly regarded in the industry, and I feel secure in their recommendations. Although past performance does not *guarantee* future results, the "Hot Picks" list for 20XX has returned 12.02% since it was published last December, significantly outperforming the S&P 500's–17.0% return. Since we began using the "Hot Picks" list in 20XX, it has yielded an average of 58.04%.*

If you are interested in learning more about the "Hot Picks" list, or if you would like to discuss possible additions to your portfolio, please contact me today! I am always available to help you realize your financial dreams.

Sincerely,

Christina Kleckner

Christina Kleckner
Financial Advisor

*As of 12/12/20XX. The S&P is an unmanaged index of 500 widely held stocks that is generally considered representative of the U.S. stock market.

▶ SYMPATHY

There are certain social responsibilities involved when a business colleague either dies, or experiences a death in the immediate family. If the colleague himself dies, it is appropriate to write his family a sympathy letter. If a close member of the colleague's family dies, then write a sympathy letter to your colleague. How well you know, or knew, the person involved has much to do with how you write your letter.

If the person was not a member of your company—but was just a business associate— a letter typed on business letterhead is appropriate. If the person was a close colleague or worked for your company, then the letter should be handwritten on card stock or personal stationery. The better you knew the person, the more personal your comments can be. If you didn't know the individual very well, your letter will come off as insincere when you go overboard with personal comments or offers of help.

The most sincere approach is to match the depth of your sympathy and support with how well you knew the person.

---

478 Beckman Road
Las Vegas, NV 89131

June 14, 20XX

Mrs. Smith
2369 North Locklear Trail
Las Vegas, NV 89131

Dear Mrs. Smith,

On behalf of everyone at Jones and Associates, I would like to express our sincere sympathy in the loss of your husband, Tom. We were deeply saddened by the news, and wish you and your family strength and peace during this very difficult time.

Very sincerely,

Burle Simpson
Burle Simpson

▶ THANK YOU

Thank you letters are written for a variety of reasons, some obvious ones including: gifts, invitations, and business referrals. In all cases, thank you letters can be brief, but they must be sincere. They should also be sent in a prompt manner, so the good deed is recognized right away.

All thank you letters must, without exception, say, "Thank you." It is better to include your thank you in the first paragraph, if not the first sentence. If the person you are writing to is a business associate, then responding with a typed thank you on company letterhead is appropriate. If the recipient is someone you know well, a handwritten letter on personal stationery is more appropriate. The better you know someone, the more personal you can make the comments in your letter.

When you get to the end of the letter, be upbeat and sincere. No need to say thank you again, after you have already said it in your first paragraph. Repeating thank you over and over again takes away from the sincerity of your message:

6895 Glendale Boulevard
Alexandria, VA 22312

February 24, 20XX

Mr. and Mrs. Kelly
7465 South Shore Drive
Alexandria, VA 22314

Dear Mr. and Mrs. Kelly,

On behalf of Yorkshire Homes, I would like to thank you for choosing us as your Custom Home builder. We are delighted to serve you in any way we can, both before completion of your home, and after you and your family have moved in.

We offer a full one-year warranty, free of charge, for all structural issues, and also on all major appliances in your new home. Please see the details of the warranty in your "New Home Owners" guide. For your convenience, there is a 24-hour customer service number that you may call with any questions, needs or concerns. If you are not completely satisfied with our service, please feel free to contact me directly at any time, and I will see that your needs are cheerfully and thoroughly met. Our friendly and professional staff is here to serve you.

Please take the time to carefully read all of the covenants and restrictions for your new development, as they are written for your safety and for the peaceful enjoyment of your new home and neighborhood.

We honor you as a valued client, and we will make it our business to ensure your family's happiness in your new home. Remember, we are only a phone call away!

Sincerely,

Vernette Dickinson
Vernette Dickinson
CEO, Yorkshire Homes

Transmittal letters function much in the same way as cover letters function for resumes. In short, they serve as a cover letter for business documents. So, if you are sending a payment or a contract, your transmittal letter should briefly explain the contents of the attachments.

Transmittal letters are short—details are all written in the enclosed or attached material.

5645 South Lee Street
Williamsburg, VA 86054

March 3, 20XX

Ms. Donna Williams
475 South Tidewater Road
New York, NY 10012

Dear Donna:

Attached is the lease agreement for 128 Monticello Drive, Williamsburg, VA 86054. Please review thoroughly to see if any changes are needed.

We can go through the formal lease and any other paperwork when you arrive. Let us know if there is anything we can do to make your transition easier. Don't hesitate to call.

We are looking forward to your arrival. Welcome to Williamsburg!

Best wishes,

Mike Worden

Mike Worden
Property Manager

► WELCOME

In business, you constantly have to look for opportunities to strengthen client relationships. When a client applies for a line of credit, makes a first purchase, or establishes a new account, you must jump at the chance to say, "Welcome!"

This serves a couple of purposes. First, it recognizes your sincere interest in your client; and second, it gives you an opportunity to cultivate a new business relationship. You must be sure to personalize your message, but keep it professional. Express your desire for a long and successful partnership, and be sure your client knows she can count on you for anything.

4750 Carthage Court
Orangevale, CA 95662

March 2, 20XX

Ms. Kimberly Denton
632 Wall Drive
Princeton, NJ 08543

Dear Kimberly,

Welcome to The Elite Financiers! It was a sincere pleasure meeting you last Friday. I was delighted to learn of your interest in kayaking, and I hope we'll have the chance to talk more about it when you return from your trip to Hawaii.

As we discussed in our meeting last week, your $10 million line of credit has been approved. You will receive full documentation within the next two weeks. Feel free to call with any questions after you look over the paperwork. I will call you on Thursday, February 14, to set up a time for the signing. Of course, we recommend that you have your team of attorneys present at the time of the signing.

In the meantime, we at The Elite Financiers want to let you know that we feel privileged to have you as a new client. The Elite Financiers serves wealthy people, like you, who have intricate financial requirements. Our goal is to allow you easy and comfortable access to Elite Financier's plentiful resources and to tailor financial solutions to fit your needs.

I have enclosed both my business and personal telephone numbers, so please don't hesitate to contact me for any banking issues. I have also enclosed all other pertinent contact names and telephone numbers for your accounts. These individuals will serve as trusted and valuable resources for all of your financial needs, and should be able to promptly answer any questions you may have as a brand new client.

I look forward to speaking with you again on February 14! Until then, enjoy your wonderful Hawaiian vacation, and please call me with any questions. We at The Elite Financiers look forward to a successful and rewarding partnership with you for many years to come.

Regards,

Jack Brown
Jack Brown

Grammar

We made too many wrong mistakes.

—YOGI BERRA

here is no way around it. Following the rules of grammar in your business writing is essential. It helps you avoid making "too many wrong mistakes." It also helps your message come through like a bolt of summer lightning, instead of a garbled display of misplaced words and poor spelling. Although no one has ever proclaimed that learning the rules of grammar is a fabulously fun thing to do, your correct usage of it *will* make a difference in the way you are perceived on paper. Just like dressing appropriately for work or using the right tone in your letters, using correct grammar sends a message to others: this person is smart, thorough, dependable, accurate, and clear.

Grammar is defined as a set of rules intended to make language make sense—rules that make it easier to communicate. In other words, grammar rules are meant to *help* you, not make you want to run for the hills. You can use grammar tools to help you make a good point, emphasize an idea, or to just send a clear message. And, the more you know about these tools, the better a writer you will be.

In this section, you will find all of the grammar basics that will help you write effective, powerful, and *correct* business correspondence. The topics include parts of speech, sentence types and structure, punctuation, capitalization, some spelling tips, and a list of commonly misspelled words.

So, here come the rules of the road—grammar rules, that is—that will help you avoid making any wrong mistakes in your business writing. Then you will be headed down a path of success.

▶ PARTS OF SPEECH

All words in the English language fit into eight neatly defined groups of words. These groups of words are called the parts of speech—adjectives, adverbs, nouns, pronouns, interjections, conjunctions, prepositions, and verbs.

ADJECTIVES

Adjectives are words that describe or modify nouns or pronouns. They add information by describing people, places, or things in a sentence. These words add spice to our writing. There are four general categories of adjectives: **descriptive, limiting, compound,** and **articles**.

Descriptive adjectives are the type most often associated with adjectives. Their purpose is to qualify the properties or behavior of nouns or pronouns. In the following examples, the descriptive adjective is italicized and the noun or pronoun being modified is underlined:

> The *new* <u>supervisor</u> met his staff today.
> The *rapid* <u>growth</u> caused problems with customer service.
> The stress of the job brings out the *real* <u>him</u>.

In most cases, the adjective comes immediately before the noun or pronoun it is modifying. However, the adjective is sometimes used in the predicate form. In these instances, it is found *after* the subject of the sentence. Here are some examples of descriptive adjectives in predicate form:

> The <u>project</u> was *hard*.
> The <u>budget</u> is *small*.
> The <u>report</u> is *inaccurate*.

Limiting adjectives place boundaries or limits on the noun or pronoun they're modifying. These limits are usually quantified by numbers, size, or time:

The *two* <u>lawyers</u> worked on the case.
The *incomplete* <u>analysis</u> forced the cancellation of the project.
The *late* <u>start</u> put the sales team behind.

It is important to distinguish between descriptive and limiting adjectives because they are often both used in the same sentence. When this is the case, always place the limiting adjective before the descriptive adjective in the sentence:

Poor form:
The *new two* <u>agents</u> went to lunch together.
The *commodity late* <u>report</u> gave much insight.
She gave him her *personal, full* <u>attention</u>.

Good form:
The *two new* <u>agents</u> went to lunch together.
The *late commodity* <u>report</u> gave much insight.
She gave him her *full, personal* <u>attention</u>.

Compound adjectives are formed by combining two or more adjectives. These new, combined words are easy to spot because they usually have hyphen between them. For example:

The <u>newly-formed</u> committee met last Monday night.
The <u>twenty-minute</u> meeting was the first of many.

The final category of adjectives is **articles**. This small collection of adjectives is limited to the words *the*, *a*, and *an*. The word *the* is called a definite article. When used in a sentence, it expressly defines *the* noun as opposed to *any* noun. The articles *a* and *an* are called indefinite articles because they indirectly refer to a noun. For example:

Joe finished *the* report. (definite)
Joe finished *a* report. (indefinite)
Marcia drafted *the* agreement. (definite)
Marcia drafted *an* agreement. (indefinite)

The two examples using definite articles clearly relate to the report or the agreement. The two indefinite article examples talk about a report or an agreement. These could be any report or agreement compared to the report or agreement described by the definite article.

Since there's only one definite article (*the*), no decision is required as to which one to use. But there are two indefinite articles (*a, an*); so how do you know which one to use?

The general rule is this: If the first letter of the noun following the article is a consonant, then use the article *a*. If the first letter of the noun is a vowel, then use the article *an*. Here are some examples:

a <u>bank</u> representative	*an* <u>advancement</u> in rank
a <u>card</u> from relatives	*an* <u>asset</u> allocation issue
a <u>downturn</u> in sales	*an* <u>excuse</u> of the worst kind
a <u>factory</u> spokesperson	*an* <u>exercise</u> in futility
a <u>home</u> equity loan	*an* <u>ideal</u> situation
a <u>joke</u> book	*an* <u>idiosyncrasy</u> of his
a <u>machine</u> repairman	*an* <u>occasion</u> for celebration
a <u>pair</u> of socks	*an* <u>oddity</u> in his behavior
a <u>stock</u> report	*an* <u>upset</u> stomach
a <u>victory</u> over defeat	*an* <u>upturn</u> in the economy

Although this general rule governing *a* and *an* works a majority of the time, there are some exceptions. For instance, the article *a* should be used for all words that start with the sounds of *h, long u* or *whuh* (as in the "whuh" sound in the word, "once"), regardless of the first letter of the word. The article *an* should be used—regardless of the first letter—where the first sound of the word is any vowel sound (except long u), or with words that start with a silent *h*.

a <u>euphemistic</u> phrase	*an* <u>FCC</u> ruling
a <u>European</u> vacation	*an* <u>HMO</u> representative
a <u>home</u> equity loan	*an* <u>LLC</u> corporation
a <u>hot</u> commodity	*an* <u>MBA</u> program
a <u>hotel</u> chain	*an* <u>NBA</u> team
a <u>once</u> great corporation	*an* <u>RJ Reynolds</u> subsidiary
a <u>one</u> track mind	*an* <u>SEC</u> regulation
a <u>one-way</u> ticket	*an* <u>x-ray</u> machine
a <u>unified</u> effort	*an* <u>herbal</u> tea
a <u>uniform</u> product	*an* <u>honorarium</u> for speaking
a <u>union</u> contract	*an* <u>hourly</u> basis

Since the job of adjectives is to describe a noun or pronoun, these words are often used for comparison. When comparing adjectives, there are a few rules to learn. First, the three levels of comparison are called: **positive, comparative**, and **superlative**. We use these tools in everyday life, so now you know the grammatical name for them.

How you modify adjectives depends on the number of syllables in the word. For example, adjectives with only one syllable are made comparative by adding *er*, and made superlative by adding *est*:

Positive	Comparative	Superlative
big	bigger	biggest
cold	colder	coldest
fast	faster	fastest
hot	hotter	hottest
late	later	latest
long	longer	longest
short	shorter	shortest
small	smaller	smallest
thin	thinner	thinnest
warm	warmer	warmest

Adjectives with two syllables can form the comparative by either adding *er* to the positive, or by placing the word *more* or *less* before the adjective. The superlative can be formed by adding *est* to the positive, or by placing the word *most* or *least* before the adjective:

Positive	Comparative	Superlative
angry	angrier	angriest
careful	more (less) careful	most (least) careful
frequent	more (less) frequent	most (least) frequent
happy	happier	happiest
hungry	hungrier	hungriest
often	more (less) often	most (least) often
patient	more (less) patient	most (least) patient
quiet	quieter	quietest
shallow	shallower	shallowest
sincere	more (less) sincere	most (least) sincere

Adjectives with three or more syllables can only use *er* to form the comparative and *est* to form the superlative:

Positive	Comparative	Superlative
advantageous	more (less) advantageous	most (least) advantageous
adventurous	more (less) adventurous	most (least) adventurous
comparable	more (less) comparable	most (least) comparable
erroneous	more (less) erroneous	most (least) erroneous
flexible	more (less) flexible	most (least) flexible
laborious	more (less) laborious	most (least) laborious
monotonous	more (less) monotonous	most (least) monotonous
pessimistic	more (less) pessimistic	most (least) pessimistic

| reluctant | more (less) reluctant | most (least) reluctant |
| tedious | more (less) tedious | most (least) tedious |

Some adjectives have irregular comparative and superlative forms:

Positive	**Comparative**	**Superlative**
Good	Better	Best
Bad	Worse	Worst

When a sentence refers to one of the five human senses—sight, hearing, touch, taste, and smell, use an adjective rather than an adverb to describe the action.

Poor form:
> The audience sounds *quietly*.
> The food tastes *oddly*.
> The material feels *warmly*.
> The contract appears *professionally*.

Good form:
> The audience sounds *quiet*.
> The food tastes *odd*.
> The material feels *warm*.
> The contract appears *professional*.

When multiple adjectives are used consecutively in a sentence to modify the same noun, one of two rules must be applied. The first rule is to read the sentence and ask the question: "Would a conjunction fit correctly between the two adjectives?" If the answer is yes, then place a comma between the two consecutive adjectives.

> Tom is an *efficient, productive* manager. (Tom is an efficient <u>and</u> productive manager).
> The CDC is an *accomplished, renowned* institution. (The CDC is an accomplished <u>and</u> renowned institution).
> The *slow, gradual* approach is always best. (The slow <u>and</u> gradual approach is always best).

The second rule concerning multiple adjectives is used when the first adjective modifies the second adjective, changing the meaning of the sentence. In these situations, do not add a comma between the consecutive adjectives:

The *quarterly shareholders* report will come out next week.
The *antique oak* furniture matched the décor of the office.
The *faded blue* color fit well with the rest of the clothing line.

ADVERBS

Adverbs are descriptive words just like adjectives. However, instead of describing nouns or pronouns, adverbs modify verbs, clauses, adjectives, and even other adverbs. Most adverbs are easily identified because the majority of them end with the suffix *-ly*. In fact, many adjectives can be converted to adverbs simply by adding *-ly*:

angry	angrily	glad	gladly	late	lately	loud	loudly
most	mostly	near	nearly	quick	quickly	quiet	quietly
slow	slowly	smart	smartly	terrible	terribly	vast	vastly

Some words end in *-ly*, but are adjectives (not adverbs) such as:

costly	daily	early	lively	lonely	monthly
neighborly	orderly	timely	weekly	worldly	yearly

In addition to these modifiers, there are two special types of adverbs: **conjunctive** and **interrogative**.

Conjunctive adverbs join thoughts and phrases:

however	nevertheless	then	therefore

Funding is important in business; *however,* there's more to it than that.
The last round of fundraising was unsuccessful; *nevertheless,* they must continue.
If the company can find the funding, *then* it can proceed with the expansion.
The funding did not come through; *therefore,* the expansion is delayed indefinitely.

Interrogative adverbs are *how, what, where, when,* and *why.* They ask questions that modify verbs, clauses, adjectives, and adverbs.

Like adjectives, adverbs are used for comparison. The same three states of comparison are used: *positive, comparative,* and *superlative.*

If an adverb has a single syllable, add -er for the comparative and -est for the superlative:

Positive	Comparative	Superlative
fast	faster	fastest
late	later	latest
quick	quicker	quickest
slow	slower	slowest
soon	sooner	soonest

If an adverb has two or more syllables, add the word *more* or *less* for the comparative and *most* or *least* for the superlative:

Positive	Comparative	Superlative
deadly	more (less) deadly	most (least) deadly
deeply	more (less) deeply	most (least) deeply
friendly	more (less) friendly	most (least) friendly
quickly	more (less) quickly	most (least) quickly
quietly	more (less) quietly	most (least) quietly
surely	more (less) surely	most (least) surely
truly	more (less) truly	most (least) truly

When using adverbs, it is best to place them as near as possible to the word or clause it is modifying. Here the adverb is italicized and the word it is modifying is underlined.

Poor form:
The supply department <u>ran out</u> of spare parts *nearly*.
We <u>must remember</u> the customers *also*.
They <u>remember</u> what they're supposed to do *scarcely*.

Good form:
The supply department *nearly* <u>ran out</u> of spare parts.
We *also* <u>must remember</u> the customers.
They *scarcely* <u>remember</u> what they're supposed to do.

Many sentences contain infinitives (verbs with the word *to* before them). Many writers make the mistake of splitting an infinitive and placing an adverb between *to* and the verb. In most cases, it is better not to split the infinitive and to place the adverb in another location.

Poor form:

Marketing wants to *aggressively* pursue these customers.

The labor talks are going to *shortly* conclude.

The company needs to *abruptly* end its hiring practices.

Good form:

Marketing wants to pursue these customers *aggressively*.

The labor talks are going to conclude *shortly*.

The company needs to end its hiring practices *abruptly*.

CONJUNCTIONS

Conjunctions are words that connect other words, phrases, and clauses. They allow us to tie together ideas and thoughts within a sentence. There are three types of conjunctions; **coordinating, correlative,** and **subordinating.** *And, but,* and *or* are the words that come to mind when most people think of conjunctions. *Yet* and *nor* are also examples of **coordinating conjunctions** For example:

Bruce *and* Patty bought a new house.

Mike went to the store, *but* it had closed.

They told him not to do it, *yet* he did it anyway.

In the first sentence, the conjunction *and* connects the two nouns Bruce and Patty. In the second two examples, *but* and *yet* connect a dependent clause to an independent clause.

Correlative conjunctions usually travel in pairs. The following combinations form correlative conjunctions: *as . . . as, both . . . and, either . . . or, neither . . . nor, not only . . . but, not so . . . as, so . . . that. For example:*

Both the teacher *and* the student were pleased with the results.

Either he goes *or* I go.

Neither the coach *nor* the players were ready for the game.

The condition of the car was *so* bad *that* it could not be salvaged.

Subordinating conjunctions are used to connect two clauses in a sentence. Words like *after, before, if, since, than, that, unless, until, when,* and *where* are just a few examples of subordinating conjunctions. For example:

They agreed to the deal *after* I lowered the price.

The shareholders are more frustrated *than* angry with management.

The company has stopped hiring new people *until* they are profitable again.

INTERJECTIONS

Interjections are words that express great emotion. They often begin a sentence. In fact, they usually stand on their own, separated from the sentence they are describing. Interjections also sometimes "borrow" words from other parts of speech. *Yes, no, right, terrific,* and *fantastic* are all examples of words that come from other parts of speech, but that serve as interjections as well.

So, remember to use interjections sparingly, as they are better reserved for conversations than business writing. In fact, of the eight parts of speech, they are probably the least often used parts of speech in business writing. The examples below are examples of phrases with interjections that we are more likely to hear in conversation than read in business writing. But, it is always a good idea to keep in mind those words and phrases that should be saved for either very informal business correspondence, or for a casual chat with a coworker.

> *Great!* I'll see you tomorrow.
> *No,* that's not how you do it.
> *Oh,* I understand now.
> *Wow!* That's a great idea.

The third sentence is an example of an interjection that is connected to the sentence it is describing.

Another common practice with interjections is the use of the exclamation point. Using the exclamation point to separate the interjection from the sentence adds emphasis or surprise to the statement. Following is an example showing the difference between separating an interjection and including the interjection in the sentence. Which one do you think has more impact?

> *Wow,* I didn't think you'd make it.
> *Wow!* I didn't think you'd make it.

The second sentence is obviously the stronger statement. As a rule, interjections don't have to be separated from a sentence.

NOUNS

Nouns account for a great deal of the words in the English language. They name ideas, objects, persons, places, and qualities. There are five types of nouns: **abstract, collective, common, compound,** and **proper.** Each has its own purpose and unique collection of words.

Abstract nouns include ideas or qualities such as *freedom, justice,* or *liberty.* Don't confuse abstract nouns with adjectives or adverbs. For example, *confusion* is a state of mind; therefore, it is an abstract noun. *Confused* describes a person's behavior or situational behavior, so it is considered an adverb.

Collective nouns are used when talking about groups of animals, people, or things. *Herd, crowd,* and *collection* are all examples of collective nouns. A common mistake is pluralizing collective nouns when they refer to a group of things. It is important to remember that the noun should be singular when talking about a group and plural when talking about more than one group. The first sentence below uses a singular form of the collective noun and the second sentence uses the pluralized form:

> A *herd of buffalo* arrived at Yellowstone.
> *Herds of buffalo* arrived at Yellowstone.

Although both sentences refer to groups of buffalo, the first sentence describes a single herd of bison, while the second sentence conjures up the image of many groups of buffalo arriving at Yellowstone.

Common nouns are words that are typically thought of as nouns—these include people, places, and things. These nouns have no special rules regarding capitalization, pluralization, or punctuation. Common nouns are just basic, run of the mill words that give our sentences the framework with which we can attach our ideas. The italicized words below are nouns that name the person, place, or thing in each sentence. For example:

> The *company* started business in 1967.
> John managed the *store* for 16 years.
> The *children* sold lemonade at the picnic.

Since common nouns can name people and places, it is possible to confuse them with proper nouns (discussed ahead). Common nouns for people and places include words like *boy, girl, kitchen,* or *house.* Proper nouns name specific people or places, like *George Washington* or *New York City.* There are different rules for using common and proper nouns, so make sure you know which one you need.

Compound nouns are formed by combining two or more words. The individual words used to build the compound noun don't have to be nouns themselves—sometimes a verb and noun are used, and sometimes an adjective and a noun are put together. Occasionally two nouns are used to form the compound noun. No matter what the individual components, once the compound noun is complete, it must be used as a noun in the structure of the sentence.

A unique quality of compound nouns is that they are spelled either as solid words (written as a single word), hyphenated words, or spaced words (written as two words separated by a space). Here are some examples of compound nouns:

Solid Words	Hyphenated Words	Spaced Words
buyback	add-on	bean counting
downturn	cross-reference	bill of lading
freelance	drive-in	data processing
phaseout	hang-up	line of credit
printout	log-out	power of attorney
rollover	once-over	standard of living

There are some special rules that apply to compound nouns. For instance, if you are writing the title of someone who holds two positions, separate them with a hyphen. A person who owns a business and serves as the manager, for example, would be referred to as an *owner-operator*. Other business terms include *secretary-treasurer* and *player-manager*.

Another hyphenation rule for compound nouns requires that you add a hyphen when using the prefix *ex-* or suffix *-elect*. The titles *ex-President*, *President-elect*, and *ex-wife* are examples of this rule.

Some nouns like *doctor, nurse, lawyer,* and *judge* are gender neutral and don't require being converted into compound nouns. *Male doctor* or *female judge* are examples of unnecessary compound nouns, unless the intent of the statement pertains to the subject's gender.

Proper nouns are those names of specific individual people, places, or things. Words like *George Bush, Arizona,* and *Microsoft* are examples of proper nouns. Notice that they are all capitalized. Always capitalize every proper noun.

Nicknames and imaginative names also should be capitalized. For instance, *the Big Apple, the Big Board,* and *Mother Goose* are all proper nouns, and should therefore be capitalized. Note that the name itself is capitalized, but the article *the* is not.

Many adjectives are derived from proper nouns. *Texas* becomes *Texan, Mexico* becomes *Mexican,* and *Orwell* becomes *Orwellian.* Note that all of the derived adjectives follow the same capitalization rule as their original proper nouns. Not all proper nouns have an adjective counterpart, but if they do, always use the adjective form when describing another noun.

Occasionally a prefix is added to a proper noun. Some examples are *mid-March, trans-Siberian Railway,* and *anti-American.* When the prefix is added to the proper noun, always use a hyphen and only capitalize the proper noun. Do not capitalize the prefix.

Finally, like every great rule of grammar, there is an exception. Some proper nouns have become common nouns and no longer require capitalization. Words like *roman numeral, watt,* and *fine china* all contain proper nouns, but have become so commonplace, modern convention has changed them to common nouns. Use an up-to-date dictionary for the latest rules about capitalizing proper nouns.

There are several rules pertaining to the use of singular or plural forms of nouns. The basic rule for pluralizing nouns is to add an *s* to the end of the word. This simple rule of adding *s* works for the vast majority of nouns.

car cars	computer computers	contract contracts	document documents
food foods	form forms	letter letters	meeting meetings
office offices	paper papers	printer printers	rule rules

Now comes the hard part: all of the exceptions to the rule of pluralizing nouns. Many nouns end in the letters *s, x, ch, sh* or *z*. Adding *es* to the singular form pluralizes these words.

dish dishes	fax faxes	match matches	sandwich sandwiches
stitch stitches	tax taxes	watch watches	wish wishes

Nouns that end in *y* are also complicated. For instance, if the letter before the *y* is a consonant (like *candy*), it is pluralized by dropping the *y* and adding *ies* (*candies*).

baby babies	city cities	company companies
factory factories	memory memories	penny pennies
secretary secretaries	subsidy subsidies	territory territories

If the letter before the *y* is a vowel, it is pluralized by adding an *s* to the end of the word.

attorney attorneys	bay bays	boy boys	day days
play plays	relay relays	toy toys	way ways

Nouns ending in *o* have their own rules. If the letter before the *o* is a vowel, pluralize the singular by adding an *s*.

portfolio portfolios	ratio ratios	scenario scenarios

If there is a consonant before the *o* at the end of the noun, the rule for pluralizing the word is more complex—it depends on the word you are using. Some words are pluralized by adding an *s*.

logo logos	memo memos	photo photos	typo typos

Some words are pluralized by adding *es*.

embargo embargoes	potato potatoes	tomato tomatoes

Some words can be pluralized by adding *s* or *es*. These are instances where either form is accepted.

cargo cargos cargoes   *domino dominos dominoes*   *no nos noes*

Another group of words that share pluralization rules are nouns that end in *f, fe* or *ff*. Most of these nouns are pluralized by adding *s*.

proof proofs   *brief briefs*   *tariff tariffs*

Some of these nouns are pluralized by dropping the *f* or *fe* and replacing it with *ves*.

calf calves   *knife knives*   *leaf leaves*   *life lives*

There are even some nouns ending in *f, fe* or *ff* that can be pluralized using *s* or *ves*.

dwarf dwarfs dwarves   *scarf scarfs scarves*   *wharf wharfs wharves*

Some nouns never require pluralization. They are always considered plural even if they refer to a single item or issue.

assets   credentials   earnings   goods   proceeds   savings   winnings

Nouns that end in *ics* are spelled the same whether they are singular or plural. But, their meaning in the sentence determines if the verb must be singular or plural. *Acoustics, economics, ethics, politics,* and *statistics* are examples of nouns ending in *ics*. If the meaning of the noun refers to a body of knowledge then the verb must be singular. If the noun refers to qualities or activities then the verb should be plural.

> The study of *statistics* is useful in business.
> The *statistics* are not favorable.

The last group of nouns is known as **irregular nouns**. They follow no uniform rule. The pluralization of these words must either be looked up or memorized.

addendum	*addenda*	*basis*	*bases*	*crisis*	*crises*	*child*	*children*
criteria	*criterion*	*foot*	*feet*	*man*	*men*	*matrix*	*matrices*
medium	*media*	*synopsis*	*synopses*	*thesis*	*theses*	*woman*	*women*

Nouns are often used to show possession of something. One basic rule with nouns and their possessive form is to add an apostrophe and an *s* to the end of a singular noun—this forms the possessive. For example:

Singular	Plural
book	book's
contract	contract's
employee	employee's
report	report's

Not all style guides are the same, but an exception to this rule applies to the possessive form of a proper noun that ends in the letter *s*. In these cases, simply add an apostrophe after the letter *s* at the end of the word. For instance:

Singular	Possessive
Jones	Jones'
Kansas	Kansas'
Louis	Louis'
Memphis	Memphis'
Texas	Texas'
Thomas	Thomas'

If the noun you are using is plural, simply add an apostrophe at the end of the word after the *s* to form the possessive:

Plural	Possessive
agreements	agreements' (. . . components)
companies	companies' (. . . investors)
employees	employees' (. . . opinions)
offices	offices' (. . . overall structure)
supervisors	supervisors' (. . . standards)

Irregular nouns not only have their own rules regarding pluralization, but they are unique in how they show possession as well. To show possession, add an apostrophe and an *s* at the end of the noun:

Singular	Plural	Possessive
child	children	children's
man	men	men's
woman	women	women's

PREPOSITIONS

Prepositions are connecting words that link a noun or pronoun to another word in a sentence. They are often used to show a relationship of space or time. For example:

> The <u>letter</u> *on* the <u>table</u> is next year's contract.
> The <u>day</u> *after* <u>tomorrow</u> is the stockholders' meeting.

The first sentence uses the preposition *on* to relate the spatial relationship between the letter and the table. The second sentence uses the preposition *after* to describe the time relationship between today and tomorrow. *On the table* and *after tomorrow* are prepositional phrases.

Here is a list of common prepositions:

aboard	*about*	*above*	*after*	*among*	*around*	*at*	*before*
behind	*below*	*beneath*	*beside*	*between*	*by*	*except*	*for*
from	*in*	*inside*	*into*	*like*	*of*	*off*	*on*
outside	*over*	*to*	*under*	*up*	*upon*	*until*	*with*
within							

Superfluous prepositions are prepositions that add nothing to the meaning of the sentence. In these cases, delete the prepositions from the sentence. Notice how the prepositions in the following sentences can be removed without changing the message.

> The construction project is almost over [with].
> The pallets of equipment are too near [to] one another.
> Where is the stapler [at]?
> The convention is now over [with].

The opposite of superfluous prepositions are **necessary prepositions**. These words are required to be in the sentence in order for it to make sense. Read the following sentences and imagine what each one would sound like without the prepositions.

> Are you going [with] me?
> What type [of] oil do you need in your car?

Another group of prepositions is used to follow certain words. These necessary prepositions are always used in combination with their respective supporting words. Below are two examples of required prepositions—the preposition is in italics and the supported word is underlined. It is important to remember that they must always be used together.

You must <u>account</u> *for* every dollar in the budget.
His report <u>consists</u> *of* several optional plans.

Here is a list of several common required prepositions:

account for	agree upon	angry with	argue about
compare to	correspond with	differ from	different than
identical to	independent of	interested in	speak with

Many times prepositions are used in the title of a book, story, or movie. As a general rule, prepositions of four letters or less are not capitalized unless they are the first word of the title. If the preposition is longer than four letters, then it is capitalized with the rest of the title.

> *Of* Mice and Men
> The Count *of* Monte Cristo
> Enemy *Among* Us

Sometimes prepositions are overused. If you see two prepositions next to one another in a sentence, chances are that one of them can probably be removed.

> **Poor form**—The birds built a nest *up under* the eaves.
> **Good form**—The birds built a nest *under* the eaves.

> **Poor form**—Everything was complete *except for* the contract.
> **Good form**—Everything was complete *except* the contract.

> **Poor form**—They started looking *outside of* the company for new candidates.
> **Good form**—They started looking *outside* the company for new candidates.

A common mistake with prepositions involves the use of *between* and *among*. Between is used when talking about two things. *Among* is used for talking about more than two things:

> The boss had to decide *between* cutting new hires or handing out Christmas bonuses.
> The work was divided evenly *among* marketing, finance, and operations.

Prepositions are often used with nouns in a series. In the series, the preposition must be used only once with the first noun of the series, or it can be used with every element of the series. Therefore, correct form is to use a preposition either once in the sentence, or before every noun:

Poor form—Cutbacks are required *in* operations, *in* personnel, administration, and management.
(*prepositions used only half the time lack parallelism*)

Good form—Cutbacks are required *in* operations, personnel, administration, and management.
Good form—Cutbacks are required *in* operations, *in* personnel, *in* administration, and *in* management.

Of all the rules governing prepositions, none is more famous than: *"No ending a sentence with a preposition."* While this rule holds true for many situations, it is not an absolute. You can still end a sentence with a preposition if it makes the sentence flow better.

Ultimately, the best technique for keeping or removing prepositions at the end of sentences is to use your ear. What would the statement sound like if you kept—or dropped—the preposition? What point are you trying to emphasize in your statement? Is this a formal statement or a casual conversational statement? This timeless question of "to keep or to cut" the dangling preposition ultimately comes down to the desired effect. Here are some examples of prepositions placed in different positions within sentences:

I thought I knew what company she worked *for.*
I thought I knew *for which* company she worked.

The first sentence sounds like a casual conversation, even though it does not strictly adhere to the rule of not ending a sentence with a preposition. But, it does sound natural. The second sentence follows the grammatical rule, but it is not the kind of statement you are likely to hear in everyday conversation. This sentence is more formal than the first, and may be appropriate in certain situations.

Many times, short questions are ended in prepositions. Here are some acceptable and unacceptable examples:

Poor form:
Is the construction project over *with*?
Where is the report *at*?
Where do you want to go *to*?

Good form:
Does he have anything to worry *about*?
What did you make it *with*?
What is the report comprised *of*?

PRONOUNS

A pronoun is a word that replaces a noun in a sentence. This allows you, the writer, to avoid repeating the same noun over and over again. Here is an example of a paragraph without pronouns:

> David is the owner and manager of two local convenience stores. David opened David's first store in 1976 when David graduated from college. The next year, David married Julie and together David and Julie opened David and Julie's downtown store.

Now here is the same paragraph with the appropriate pronouns:

> David is the owner and manager of two local convenience stores. He opened his first store in 1976 when he graduated from college. The next year, David married Julie, and together they opened their downtown store.

Obviously, the second example reads and sounds better than the first. This is an excellent example of the importance of pronouns.

There are six different types of pronouns: **demonstrative, indefinite, interrogative, personal, reflexive,** and **relative.**

The words *this, these, that,* and *those* are known as demonstrative pronouns. These pronouns are used to demonstrate distance and describe either singular or plural nouns:

	Close	Far
Singular	This	That
Plural	These	Those

Take *this* computer and move it to *that* desk.
Put *these* supplies in *those* bins.

Indefinite pronouns are used to describe a person or thing, but are not specific. Here is a list of some common indefinite pronouns:

all	another	any	anybody	anyone	anything	anywhere
both	each	either	every	few	little	many
more	most	much	neither	nobody	none	no one
nothing	nowhere	one	others	several	some such thing	

Some examples in sentence form:

> *Anyone* wishing to attend the Christmas party must buy tickets today.
> *No one* answered the phone all day.
> *Few* realize the gravity of the situation.

In the examples above, the indefinite pronouns took the place of nouns in the sentences. However, there are times when indefinite pronouns act as adjectives. Remember the difference between pronouns and adjectives. A word is considered a pronoun when it replaces a noun in a sentence. A word is considered an adjective when it describes a noun in a sentence. For example:

> *Many employees* didn't understand the new policy.
> *Many* didn't understand the new policy.

In the first sentence, *many* is considered an adjective because it describes the noun, *employees*. In the second sentence, *many* is considered an indefinite pronoun, as it takes the place of the word *employees*.

The next group of pronouns is called **interrogative pronouns**. These words ask questions. They include:

what   which   who   whoever   whom   whomever   whose

Examples in sentences:

> *What* does that have to do with anything?
> *Who* did you say was supposed to be at this meeting?
> To *whom* do we owe the honor?

There is one particularly important rule to remember about interrogative pronouns: it is the rule defining when you should use the words *who* or *whom*. The word *who* is the nominative form and the word *whom* is the objective form of the same pronoun. If the pronoun can be rewritten using other pronouns like *he, she, I* or *we* to answer the question, then *who* should be used. If the sentence can be rewritten using *him, her, me,* or *us* to answer the question, then use *whom*:

> *Who* is hosting the meeting?
> *She* is hosting the meeting.

> *Who* wrote the annual report?
> *He* wrote the annual report.

Whom did you talk to in management?
You talked to *him*.

To *whom* did *she* give the recommendation?
She gave it to *me*.

Probably the most widely recognized forms of pronouns are **personal pronouns**. These are words used to replace nouns pertaining to people. Personal pronouns are categorized in two ways: by voice and by case. The three voices of pronouns are **first person, second person,** and **third person**. The three categories of case are: **nominative, objective,** and **possessive**.

This table helps organize the personal pronouns by voice and by case:

Voice	Nominative	Objective	Possessive
first person singular	*I*	*me*	*my / mine*
first person plural	*we*	*us*	*our / ours*
second person singular	*you*	*you*	*your / yours*
second person plural	*you*	*you*	*your / yours*
third person singular masculine	*he*	*him*	*his*
third person singular feminine	*she*	*her*	*her / hers*
third person singular neutral	*it*	*it*	*its*
third person plural	*they*	*them*	*their / theirs*

There are some basic guidelines for using personal pronouns. First, choosing the right voice is imperative. Writing in the *first person* used to be reserved only for those letters and memos sent to someone you know very well—this style was often deemed inappropriate in business writing. In today's business correspondence, however, using the first person has not only become acceptable, but it is now the more preferred, comfortable style of business writing.

The *second person* is still a common voice used in business correspondence. It presents enough formality in the tone of the message to be appropriate in both business letters and memos. The *third person* can also be used in business communications. However, it is often mixed with the second person voice. If the third person is used exclusively, you run the risk of becoming too stilted in style. Here are some examples of sentences using the three voices:

Poor form:
It appears as though the meeting did not go very well. (third person singular)

Poor form:
You must realize that the meeting did not go very well. (second person singular)

Good form:
I do not think the meeting went very well. (first person singular)

The next set of rules guiding personal pronoun use deals with case. The **nominative form** is used if the pronoun is the subject of the verb. For example:

I directed the marketing strategy segment of the meeting at last night's conference.
You were the person responsible for the department's turn-around.
He wrote the report that was sent to the customer.
They decided not to expand operations at this time.

The other rule governing the use of the nominative personal pronouns is when a form of the verb *to be* (*am, is, are, was, were*) is used in a sentence. In these cases, if the pronoun follows the verb *to be*, it should be in the nominative form. The examples below show the verb underlined and the pronoun in italics.

That <u>is</u> something *I* would say.
Who <u>are</u> *we* to judge?
Those <u>were</u> accounts assigned to *you*.
This <u>is</u> an example where *he* should have spoken up.
The contracts <u>were</u> a responsibility *they* should have handled.

The **objective form** of pronouns includes *her, him, it, me, them, us,* and *you*. There are three instances that require the use of the objective form of pronouns.
The first is when the pronoun is the direct or indirect object of the verb.

My supervisor gave *us* the day off.
The gold watch was a present for *you*.
Patience is not one of *her* qualities.
He told *them* the account was overdue.

The second is if the pronoun is the subject or object of an infinitive. Here the infinitive is underlined and the pronoun is italicized.

The chairman requested *me* <u>to attend</u> the conference.
The finance department is the place for *you* <u>to start</u> a career.

The new account will give *him* something <u>to do</u>.
The budget is something for *them* <u>to decide</u>.

The third rule for using the objective form is if the pronoun is an object of a preposition. In these examples, the preposition is underlined and the pronoun is italicized.

That is an issue <u>between</u> *you* and *me.*
The rest of the staff is <u>different from</u> *him.*
The accounting statistics must <u>agree with</u> *them.*

The **possessive form** of pronouns is used to indicate possession.

The merger was *my* idea.
It was *their* moment in the sun.
Our decision was to drop the matter.

Like the nominative and objective forms of personal pronouns, possessive pronouns have many rules. Some of the most important ones follow.

First, if the pronoun is just before the noun it modifies, the correct possessive pronouns are: *my, your, his, her, its, our,* and *their.*

My report was turned in late.
Her attitude started to affect her work.
Their car broke down on the way to the office.

Next, if the pronoun is not adjacent to the noun it is modifying, the correct possessive pronouns are: *mine, yours, his, hers, its, ours,* and *theirs.*

The phone call you missed was *mine.*
The new account is all *yours.*
This is mine and those are *theirs.*

A third rule is if the noun being modified is a **gerund** (a word that has an *-ing* added to the end of it), then the correct form of pronoun is the possessive.

The staff was entertained by his colorful speaking.
Her incessant novel reading left her little in the way of a social life.
Huge profits were realized because of his marketing.

The final rule regarding the use of possessive pronouns deals with confusing the possessive form with contractions. The best way to determine if you should use a pronoun or a contraction is to expand the contraction and substitute it into the sentence. If the sentence makes sense, then use the contraction. If not, then use the corresponding possessive pronoun. Here are some commonly confused possessive pronouns and contractions:

Possessive Pronoun	Contraction
its	*it's (it is)*
their	*they're (they are)*
theirs	*there's (there is)*
whose	*who's (who is)*
your	*you are (you are)*

The company is expanding *its* sales territory.
It's time to go home.

Their company is located downtown.
They're the kind of people you can count on.

Theirs is a special relationship.
There's something special about that company.

Whose turn is it?
Who's going to tell the boss?

Your office is near mine.
You're taking a short cut.

Reflexive pronouns are easily identified because they all end in the suffix *-self* or *-selves*. The following words are categorized as reflexive pronouns:

Singular	Plural
Herself	Ourselves
Himself	Themselves
Itself	Yourselves
Myself	
Oneself	
Yourself	

He changed *himself* into a new person.
She checked *herself* into a hotel.
They talked among *themselves*.

Some reflexive pronouns are used in a sentence to add emphasis. When this is done, the pronouns are called intensive pronouns.

> *He* presented the case *himself.*
> *You yourself* said it couldn't be done.
> *They themselves* agree that the contract is sound and accurate.

Relative Pronouns

The list of relative pronouns looks very similar to the list of interrogative pronouns. Relative pronouns include:

who   whose   whom   which   that

The difference between interrogative and relative pronouns is how they are used. Remember, interrogative pronouns "ask questions" but relative pronouns are used to describe the connection between a main clause and a dependent clause. For example:

> *Who* is going to travel to Boston? (interrogative form)
> The boss decided *who* is traveling to Boston. (relative form)
> *Which* department needs the memo? (interrogative form)
> The supervisor determined *which* person to promote. (relative form)

An important rule of relative pronouns deals with the use of *who* and *that.* When the noun being referenced is an individual or single entity, use *who.* If the pronoun refers to a group or type of people, use the relative pronoun *that.*

> The president is the only person *who* can approve the request. (singular)
> She is the type of person *who* gets things right the first time. (singular)
> They are the kind of team *that* gets results. (plural)
> The employees saw to it *that* she got a warm send off. (plural)

If the noun being replaced is an animal, object, or place, you must use *which* or *that.* This is another important rule of relative pronouns pertaining to connecting clauses—after all, this is the purpose of relative pronouns. If the clause being introduced by the pronoun is an essential clause, use *that.* If the clause is a nonessential type, use *which.*

> Essential Clause:
> The memo *that* went out last week was incorrect.
> The labor negotiations *that* started yesterday went into the night.

Nonessential Clause:

The report, *which* took all night to complete, was a work of art.

The office furniture, *which* was recently moved out of the west wing of the building, was set up in the south wing on Monday.

One more rule of thumb about relative pronouns is to remember to place the pronoun immediately after the noun it is referencing.

Poor form:

The memo was incorrect *that* went out last week.

The labor negotiations went into the night *that* started last night.

Good form:

The memo *that* went out last week was incorrect.

The labor negotiations *that* started yesterday went into the night.

The concept of **agreement** is crucial to understanding and using pronouns. Pronouns must agree with their antecedent (the word they are replacing) in three ways: number, gender, and person. In these examples, the pronoun is italicized and the antecedent is underlined.

The <u>employees</u> will receive *their* paychecks tomorrow.

The <u>president</u> said *he* could attend the conference.

The <u>committee</u> will decide how to spend *its* resources.

<u>I</u> came to the conclusion it was time for *me* to go.

If two nouns are connected by the conjunction *and*, you must reference them with a plural pronoun.

<u>Mark</u> and <u>I</u> send *our* heartfelt congratulations.

<u>You</u> and the <u>supervisor</u> were *your* own worst enemy.

<u>Steve</u> and <u>Kim</u> discussed *their* plan.

If two singular nouns are joined by *or* or *nor*, a singular pronoun must be used. If the nouns are plural, then a plural pronoun must be used.

Singular:

Either <u>Tom</u> or <u>Mike</u> will get *his* chance to run the company.

Neither <u>Tom</u> nor <u>Mike</u> thinks *he* has a chance at promotion.

Neither <u>Trish</u> nor <u>Laura</u> considers this issue *her* responsibility.

Plural:

> Neither the <u>staff</u> nor <u>I</u> would have made that decision if it were up to *us.*
> Both the <u>writers</u> and the <u>directors</u> were happy with the way *they* performed.
> Either the <u>company</u> or the <u>union</u> will have to compromise *their* position.

VERBS

Verbs are words that depict action or a state of being. They tell the reader what is happening to the subject of the sentence, and in so doing form the core of the written language. There are three basic types of verbs: **transitive, intransitive,** and **helping** (or auxiliary) verbs. **Transitive** verbs must be linked to the object of a sentence. The verb and object are directly related and require one another's presence to complete a sentence. Here, the transitive verb is italicized and the object is underlined:

> The marketing department *published* the new <u>advertisement</u>.
> The company *reorganized* the <u>payroll</u>.
> Initial public offerings *offer* tremendous <u>opportunities</u>.

Intransitive verbs do *not* require an object to complete the sentence.

> Their strategy *changed.*
> The Internet startup *failed.*
> The project *was completed.*

Helping, or **auxiliary, verbs** are used with the past or present participle forms of other verbs. Here, the helping verb is italicized and the verb being "helped" is underlined.

> The accounts *were* <u>dropped</u> after two years of inactivity.
> The market *is* <u>panicking</u> after the interest rate hikes.
> The customers for their product *are* <u>swelling</u> in numbers.
> The new employee *has* <u>arrived</u> late for the last time.

The most common **helping** verbs are:

can	could
do	did
has	had
have	may
is	are
might	must
shall	should
was	were
will	would

All verbs have four principal parts from which all other forms are derived. These four forms are: **past**, **present**, **past participle**, and **present participle**. For most verbs, the past and past participle are formed by adding *ed* to the end of the present form. Similarly, the present participle is created by adding *ing* to the end of the present form.

Present	**Past**	**Past Participle**	**Present Participle**
borrow	borrowed	borrowed	borrowing
gain	gained	gained	gaining
measure	measured	measured	measuring
ski	skied	skied	skiing
track	tracked	tracked	tracking

Some verbs, however, have irregular forms when converted to the past, past participle, or present participle forms.

Present	**Past**	**Past Participle**	**Present Participle**
do	did	done	doing
get	got	got	getting
go	went	gone	going
see	saw	seen	seeing
say	said	said	saying

Since verbs are the most complex part of speech, they have many rules. The most important consideration when using verbs is tense. Tense gives the verb a reference of time: *past*, *present*, or *future*. It also describes what has happened, or what is going to happen. There are a total of 12 tenses: **past**, **present**, **future**, **past perfect**, **present perfect**, **future perfect**, **past progressive**, **present progressive**, **future progressive**, **past perfect progressive**, **present perfect progressive**, and **future perfect progressive**. All 12 tenses are derived from the four principal parts of a verb mentioned above.

The **past** tense is derived from the <u>past</u> part of the verb—no helping verbs are used in the past tense. Be sure *not* to use the past participle form of the verb when expressing something in the past.

    Poor form:
        I *gone* to the store.
        I *done* what was necessary.
        I *seen* what you did.
    Good form:
        I *went* to the store.
        I *did* what was necessary.
        I *saw* what you did.

The **present** and **future** tenses are derived from the <u>present</u> part of the verb. Using the present form of the verb in the present tense is straightforward; simply use the present form of the verb. If you want to express a verb in the future tense, add *will* or *shall* before the present form of the verb.

<u>Present Tense:</u>
> They *meet* all deadlines on time.
> Marketing *designs* the product catalog.
> Operations *produces* 10,000 units per day.

<u>Future Tense:</u>
> They *will meet* all deadlines on time.
> Marketing *shall design* the product catalog.
> Operations *will produce* 10,000 units per day.

The **past perfect**, **present perfect**, and **future perfect** tenses are all derived from the <u>past participle</u> part of the verb. Since the perfect tenses are derived from the past participle form of the verb, helping verbs must be used in these sentences.

The **Past Perfect** tense is used to describe something that began and was completed in the past. The helping verb *had* always comes before the past participle form of the verb.

> They *had dropped* the accounts after two years of inactivity.
> The company *had investigated* all the new hires.
> The stock *had gained* six points.

The **Present Perfect** tense indicates that something began in the past and was recently completed or is still occurring. The helping verbs *has* or *have* are inserted before the past participle form of the verb.

> They *have dropped* the accounts after two years of inactivity.
> The company *has investigated* all the new hires.
> The stock *has gained* six points.

The **Future Perfect** tense shows something that will be completed sometime in the future. The helping verbs *shall have* or *will have* accompany the past participle.

> The bus *will have arrived* already by 6 P.M.
> The company *will have moved* by April.
> The stock *will have gained* six points.

A common mistake with using the past participle form of verbs is to use the past tense form instead.

Poor form:
I *have spoke* to the management.
He *has chose* to write the report himself.
They *will have saw* fifty applications by the end of the day.

Good form:
I *have spoken* to the management.
He *has chosen* to write the report himself.
They *will have seen* fifty applications by the end of the day.

The **past progressive, present progressive, future progressive, past perfect progressive, present perfect progressive,** and **future perfect progressive** tenses are derived from the <u>present participle</u> part of the verb. Like the perfect tenses, these verbs must also be accompanied by a helping verb.

The **Past Progressive** tense deals with some ongoing action that occurred in the past. The helping verbs *was* or *were* are added to the present participle form of the verb.

She *was saying* no to the offer.
They *were tracking* the progress.
The company *was going* in the wrong direction.

The Present Progressive tense refers to an action still in progress. The helping verbs *am, is,* or *are* are added to the present participle form of the verb.

She *is saying* no to the offer.
They *are tracking* the progress.
The company *is going* in the wrong direction.

The Future Progressive tense is used to describe some future action. The helping verbs *shall be* or *will be* are added to the present participle form of the verb.

She *will be saying* no to the offer.
They *shall be tracking* the progress.
The company *will be going* in the wrong direction.

The **Past Perfect Progressive** tense is similar to the past perfect tense except that it gives a sense of continuous action. These verbs are formed by combining the helping verb *had been* with the present participle form of the verb.

They *had been working* on the government project for months.
She *had been walking* for half an hour.
He *had been telling* them no all along.

The **Present Perfect Progressive** tense is similar to the present perfect tense except that it gives a sense of continuous action. These verbs are formed by combining the helping verbs *has been* or *have been* with the present participle form of the verb.

They *have been working* on the government project for months.
She *has been walking* for half an hour.
He *has been telling* them no all along.

The **Future Perfect Progressive** tense is similar to the future perfect tense except that it gives a sense of continuous action. These verbs are formed by combining the helping verbs *shall have been* or *will have been* with the present participle form of the verb.

They *will have been working* on the government project for months.
She *will have been walking* for half an hour.
He *will have been telling* them no all along.

▶ SENTENCES

A sentence is a complete thought, with the subject or subjects clearly looking forward to the verb or verbs telling us what the subject is up to. Trouble comes when we move away from the simple sentences of our youth to compound and complex structures. We go from "The cat ran," to "The cat, a sleek black beauty spitting and snarling, dodged the safari group and stealthily ran through a mass of brambles and thick brush to quiet safety."

At this point, the trouble arrives in the form of run-ons or fragments, or the possible misuse of commas and periods. If you pay attention to the basic rules of sentence structure, you will be able to avoid the dreaded fragments and run-ons, thereby presenting yourself in the most professional light possible.

SIMPLE AND COMPOUND SENTENCES

A **simple sentence** contains one independent clause, which is a single subject and a single verb or verb phrase.

Mrs. Randolph gave the student a pass to class. (*Mrs. Randolph* is the single subject, and *gave* is the single verb.)

A **compound sentence** contains two independent clauses that are closely related and is usually joined by a conjunction *(and, or, nor, but, for, yet, so)*. You must put a comma after the first clause and before the conjunction.

Sacramento is the capital of California, so its local economy focuses on keeping political and business interests actively involved with one another.

Sacramento is the capital of California is one clause that could stand alone. It is joined by *so* to attach itself to the second independent clause . . . *its local economy focuses on keeping political and business interests actively involved with one another.*

COMPLEX AND COMPOUND/COMPLEX SENTENCES

A **complex sentence** contains one independent clause and one or more dependent clauses.

Kim fired the employee when his absences became excessive.

"Kim fired the employee" is a simple sentence, an independent clause. *" . . . when his absences became excessive"* cannot stand alone as an independent clause because of the word *when*. Remember, the key to writing a complete sentence is to ask yourself if it makes sense. Picture yourself saying this to someone: *"When his absences became excessive."*

Without an independent clause that explains your meaning, your expression makes no sense, so the dependent clause always "depends" on the independent to help it make sense.

A **compound/complex sentence** contains two or more independent clauses and at least one dependent clause.

When the hiring freeze was over, Chet called Matt, and Jeff yelled, "You are hired!" in the background.

"When the hiring freeze was over" is a dependent clause, making no sense if used alone. *"Chet called Matt and Jeff yelled, 'You are hired!' in the background"* are two independent clauses.

SENTENCE FRAGMENTS AND RUN-ON SENTENCES

Sentence fragments are parts of sentences, not expressing a complete thought, and often lacking a subject or a verb. They typically occur when you are following up on a thought, but close your thought too soon, leaving the rest of your idea hanging out as a fragment. You can correct the fragment by either joining it to an independent clause, or by rephrasing it to make it an independent clause on its own.

Following are several examples that explain how to incorporate different circumstances that come up in sentence formation.

Poor form:

We went to the conference in Los Angeles. After the meeting in San Diego. (*After the meeting in San Diego* makes no sense when used alone, so it should not be written as a complete sentence.)

Good form:

We went to the conference in Los Angeles after the meeting in San Diego.

Poor form:

Deborah had just met Mr. Brizgy. The new surgery center chief surgeon from Europe. (*The new surgery center chief surgeon from Europe* makes no sense when used alone. Correct it by using a comma and the words that identify Mr. Brizgy. This restatement is called an **appositive**.)

Good form:

Deborah had just met Mr. Brizgy, the new surgery center chief surgeon from Europe.

Poor form:

Mike worked all day Saturday. Finishing the annual budget right before the board meeting that evening. (*Finishing the annual budget right before the board meeting that evening* needs to be incorporated into the sentence to make sense.)

Good form:

Mike worked all day Saturday, finishing the annual budget right before the board meeting that evening.

Poor form:

Jack came to the meeting without his briefcase. Which was so typical of him. (This is a situation in which the clearest repair of the fragment *Which was so typical of him* is to create two sentences.)

Good form:

Jack came to the meeting without his briefcase. This act was so typical of him.

Run-on sentences occur when you join sentences without putting a punctuation mark in between, or when you simply join them by putting a comma between them. They can be corrected by using a semicolon, a comma, and a conjunction, or by making two complete sentences, each closed with a period.

Poor form:

Running a large corporation is a lot of work, you have to be focused and energetic.

(Notice that *Running a large corporation is a lot of work* and *. . . you have to be focused and energetic* are each able to stand alone as independent clauses. They each make sense. If you join such complete thoughts with a simple comma, then you have created a run-on, and also the impression that you don't take care with your writing.)

Good form:

Running a large corporation is a lot of work; you have to be focused and energetic.

(*You may use a semicolon to correct the run-on.*)

OR

Running a large corporation is a lot of work, *so* you have to be focused and energetic.

(*You may use a comma and a conjunction.*)

OR

Running a large corporation is a lot of work. You have to be focused and energetic.

(*You may make two complete sentences.*)

▶ PUNCTUATION

Punctuation is the use of a variety of marks in the written English language. It allows you to convey certain tones and inflections on paper that might otherwise be lost or misinterpreted. In essence, punctuation allows you to send a message without using your voice or your body language. It tells your reader whether you are excited, happy, angry, or just writing a matter-of-fact statement. If you punctuate effectively, your writing can seem almost as though you are right there in person, which is the desired effect of most business writing.

Restructuring the punctuation in a sentence can alter its meaning as easily as changing the actual words in the sentence. Using proper punctuation gives a sentence emphasis where it is needed, and also separates longer sentences into more easily defined and understood segments. There are dozens of different punctuation marks in the English language, but those covered in this section are the ones most often used in business today.

AMPERSAND (&)

The Ampersand, or "and" symbol, is often used in place of spelling out the word "and" in abbreviations, names, and titles. Usually a single space is used before and after an ampersand, unless it is part of an abbreviation where it will have no space before or after each letter.

Anderson & McKay Inc.   Atchison, Topeka & Santa Fe
B&C Enterprises   JK&P Consolidated

Another rule for using the ampersand is to use it in a long title or series. Many companies, particularly law firms and other partnerships, list the names of the primary partners. When this is the case, do not use a comma before the ampersand in the series.

Ammerman, Aguilera, Hutchison & McKinely
Hanover, Biggham, Palmer & Associates

One last rule of thumb is to always use the word *and* in a company name or title if you are not certain if it contains an ampersand or spells out the word *and*.

APOSTROPHES (')
Apostrophes are used to indicate ownership, and have been accepted as proper punctuation for the last several hundred years.

1. One general rule with apostrophes is to add *'s* to form the singular possessive.
   The *dog's* dish is empty.
   The *quarterback's* pass was intercepted.
   The *lawyer's* brief was masterfully prepared.
   Travis's book was on his desk.
   Mr. Mote's classroom is always a fun place to be.
   Chris's toy airplane flew over the house.

   *Note that the rule applies even when the singular noun ends in *s*, such as Travis and *Chris*.

2. A few plurals, not ending in *s*, also form the possessive by adding *'s*.

   The *children's* dinner was on the table.
   The *men's* room was closed.
   The *sheep's* wool was soft and fine.
   Women's rights were achieved by the active work of many.

3. Possessive plural nouns already ending in *s* need only the apostrophe added.

   The *drivers'* complaints were sent to headquarters.
   The *ships'* officers saluted each other across the open water.
   Citizens' rights are guaranteed by the first amendment.
   Five *carpenters'* toolboxes were left at the building site.

4. Indefinite pronouns show ownership by the simple addition of the apostrophe.

> *Everyone's* coats were in the hall.
> *Somebody's* credit card fell to the sidewalk.
> *Anyone's* doughnut could have been the one that stained the carpet.
> *Anybody's* car left in the No Parking zone will be towed.

5. Possessive pronouns never have apostrophes, even though some may end in *s*.

> *Our* family farm is a long time dream.
> *Her* pets are well cared for.
> *Yours* are not.
> *His* driving is erratic.

6. Use an *'s* to form the plurals of letters, figures, and numbers used as words, as well as certain expressions of time and money. The expressions of time and money do not indicate ownership in the usual sense.

> Two *s's* must be added to the client's name.
> Four *0's* followed the dollar amount on the check.
> You have too many *becauses's* in that sentence.
> These *5x7's* will be ready by Friday.
> We want to take a *week's* vacation.
> That aspirin will give you about three *hours'* relief from pain.
> After four *months'* work, the novel was near completion.

7. Show possession in the last word when using names of organizations and businesses, in hyphenated words, and in joint ownership.

> *Ben and Caryn's* company logo was a gavel and a textbook.
> I encased my *great-grandmother's* locket in a special box frame.
> The meeting of the *Organization of American States'* members took place in February.

Apostrophes may also be used to form contractions by taking the place of a missing letter or number. These contractions—shortened forms of common expressions—are acceptable both in spoken English and in informal writing. Do not use contractions in highly formal written presentations:

> *We're* (we are) ready to launch a new line of cosmetics.
> *She's* (she is) going to be on the next flight.
> *It's* (it is) never going to work the way you have (you have) designed it.

They're (they are) the ones responsible for the construction error.

The class of *'61* (1961) held their reunion in Washington.

BRACKETS ([ ])

Brackets have few common uses. They indicate your own words inserted or substituted as a way of clarification within a quotation from someone else:

> When giving a news report on the vice-president's whereabouts, the Channel 9 anchorman said, "His [Dick Cheney's] current location is unknown to us at this time."

Brackets also set off interrupting elements occurring in material already enclosed by parentheses:

> The basketball team's losses (their record was 0 and 12 [four losses at home]) were the cause of the new coach being fired.

COMMAS (,)

The rules for comma use are seemingly endless, with ten rules providing the basics for their successful management. Many times their use is optional, and most certainly their use or misuse will not cause you dire harm. Using them correctly, however, just may convince your readers that you care enough about them to be careful with your writing. There are five major areas of comma use will help you to minimize the confusion. Use commas:

- before *and, but, for, or, nor, yet, still,* when joining independent clauses.
- between all terms in a series, including the last two.
- to set off parenthetical openers and afterthoughts.
- before and after parenthetical insertions.
- in dates, titles, and quotations.

1. Use commas to separate two sentences joined by a coordinating conjunction.

    > Winston Mays won the company award, *and* he later went on to become CEO.
    > Erika fixed the copier, *but* it broke again that same afternoon.
    > Either you honor the contract, *or* the company will be forced to take action.
    > I have written the entire contract, *but* the attorneys must approve it.

2. Use commas to separate words and word groups in a series of three or more.

    > *Brian, Robin,* and *Annelise* met with the auditors to review the balance sheet.
    > The Board of Directors will *interview the staff, compile the information, and then present it to the stockholders.*

Jillian presented a *crisp, concise, well-spoken report* to the class.
Who was the *tall, strong-looking, older woman* in the back of the room?
I came, I saw, I conquered. (Julius Caesar)

3. Use commas to separate two adjectives when the word *and* could be inserted between them.

   He is a *strong, belligerent* competitor.
   He is a strong *and* belligerent competitor.

   *Do not use a comma if *and* cannot be used to separate the adjectives:

   We will hold our staff meeting at an expensive summer resort.
   (You could not say, "*expensive and summer*" resort.)

4. Use commas to set off words that are not part of the main structure of the sentence, such as introductory elements and expressions that interrupt the flow of the sentence.

   For several reasons, they moved the meeting to Montreal.
   Being of sound mind, he fired the accounting firm that was under investigation.
   As hard as he tried, he simply could not do it.

   *Some introductory elements are as simple as words of emphasis:

   Yes, I do need that report immediately!
   Now, where are you headed with that line of thinking?
   Inside, the meeting was getting more heated.

   *Clauses that start with *though* or *although* are also set off by commas:

   Although the café was closed, we peeked in the windows longingly.
    We always enjoy traveling to Maine, *though we seldom stay long enough*.

   *When starting a sentence with a weak clause, use a comma after it:

   If you have any questions about this, ask the personnel office.

   *Expressions that interrupt flow can be parenthetical in nature, or simple afterthoughts:

   I am, *as you have probably noticed*, very upset about this.
   Erin will be admitted, *I think*, after she passes the screening procedure.

5. Use commas to set apart someone or something that has been insufficiently identified.

> Lori, *the COO of our insurance rival,* is my close friend.
> That man, *who definitely knows something,* is not talking.
> Patty, *who walks with a limp,* sued the building contractor.

6. Use commas to surround the name or title of a person directly addressed.

> Will you, *Hilde,* take on the directorship of the landscape committee?
> Yes, *Doctor Gershweir,* I will.

7. Use commas to separate the day of the month from the year and after the year.

> *December 7, 1941,* is a day that will always be remembered in American history.

> *If any part of the date is left out, omit the comma:

> *December 1941* is an infamous month in American history.

8. Use a comma to separate the city from the state and after the state. This general rule is currently being modified in some businesses to exclude the use of the second comma.

> Working in *Sacramento, California,* has been a wonderful experience.

9. Use commas to surround degrees or titles used with names.

> *Leslie Worden, Ph.D,* got a job in the city she preferred.
> *David Pfaff, Vice President of Marketing,* would be proud to see the work done in his name.

10. Use a comma to introduce quotations. Remember to put the comma inside all quotation marks.

> "*Hiring him was a huge mistake,*" Dr. Hannah Rose stated.
> "*Why,*" the boss asked Stewart, "*are you always the last to leave the lab?*"

COLONS (:)

There are four basic rules for colon use: One signals the reader that something important is to follow in list format, one is used after a greeting in a business letter, one is placed between the numerals that indicate hours and minutes, and one introduces a long or formal quotation.

1. Use a colon to introduce a list when you **don't** use the words *for example* or *such as.*

   The firm ordered its usual supplies: magnets, flasks, and beakers.
   Our invoice needs to show the following: bills of lading, dock fees, and telephone charges.

   *Do not use a colon after a verb or a preposition.

   Our usual supplies *are:* magnets, flasks, and beakers. (wrong)
   We will serve our customers *with:* diligence, speed, and respect.(wrong)

2. Use a colon to follow the greeting in a business letter.

   Dear Sir: Gentlemen: Dear General Worden:

3. Use a colon to separate the numerals of hours and minutes.

   5:20 A.M.   4:30 P.S.T.   the 1:00 P.M. flight

4. Use a colon to introduce a long or formal quotation.

   In our company newsletter, CEO Olivia Churchill reported: "We were gratified to discover that the financial balance sheet reflects all your hard work and dedication. Profits have risen by an astounding 26%, supporting the fine showing in the stock market. All of us will benefit from the effort that has been demonstrated this fiscal year."

EM-DASH (—)

The em-dash essentially does the same job as the parentheses, but in a much lighter handed manner. Em-dashes indicate the same separation that is greater than commas, working effectively as the markers of abrupt change of thought or structure in a sentence. Remember to place two consecutive em-dash marks together for correct punctuation form.

   He wanted to go, whether or not he was invited—the meeting was too important to miss.
   Michael Jordan—basketball legend, golfer, baseball star—regularly made the cover of news magazines.

   The em-dash may also define a word if the context is not sufficient for the reader to gain the meaning of the word.

   The barracoon—a place where slaves were held for sale in Africa—was burned to the ground.

A single dash may emphasize an added comment.

Cade could be the CEO of the company—provided he finishes his advanced degree.

Punctuation for the dash depends on the phrasing of the interrupter phrase. If you are asking a question or making an exclamation, you must add appropriate punctuation marks.

The Pentagon—can you believe it is still standing?—continues to function as a primary planning place for the military.

ELLIPSIS (. . .)

The ellipsis mark is three spaced periods used when omitting something from a quotation. Do not use them when using your own words in a writing piece or in place of a dash. There are several rules that cover different circumstances that occur depending on where the omitted piece falls in the quotation.

1. If you omit words from the middle of a quotation, mark the omission by using three spaced periods.

   Mr. Jones, *standing rigidly in front of the hostile crowd*, replied bitterly to the question posed by the reporter.
   Mr. Jones . . . replied bitterly to the question posed by the reporter.

2. If you omit the end of the quotation, simply add a fourth period.

   Barry Brown ended the presentation with the ringing words, "*This will mark the beginning of the end. . . .*"
   Barry Brown was actually heard to say, "This will mark the beginning of the end *for corporate America.*"

   *Note the subtle difference between period placements in each sentence above. In the first sentence, there is a space between the word end and the following three periods. In the second sentence, since the quoted piece is a complete sentence followed by more text, the first of the four periods is placed immediately after the sentence.

3. If you omit words that follow a completed sentence, add the ellipsis mark to the period already there.

   Cathy Dafoe began her presentation by stating, "Sadly, what you are about to hear from me will simply sustain your current fears. . . . "

Cathy actually said, "Sadly, what you are about to hear from me will simply sustain your current fears. *Our corporate profits have plummeted at an unprecedented rate and we are facing bankruptcy.*"

EXCLAMATION POINTS (!)

1. Reserve the exclamation point for use in sentences that express very strong feelings.

    I'll expect you in Dallas on Monday, or I'll see your resignation on my desk!

2. Use the exclamation point after emphatic interjections.

    At long last! I finally get to meet the famous Buster Kirschman.

HYPHENS (-)

The hyphen is used in several different ways. It is used to indicate the division of a word at the end of a sentence, to separate spelled fractions and compound words and numbers, and to indicate prefixes in nouns and adjectives. The correct form for hyphens is just one single hyphen mark.

1. When separating words at the end of a line, be sure to hyphenate at the end of the appropriate syllable. Do not separate words with only one syllable.

    Your flight will take you through Atlanta on your way to the *meeting* in Miami.
    How many people will be hired in the wake of the *company's* need for delivery personnel?

2. Use the hyphen in compound numbers from twenty-one to ninety-nine and in fractions that are used as adjectives.

    I am *twenty-six* years old.
    Olivia was elected the *forty-first* President of the United States.
    The Republicans won by a *two-thirds* majority.(adjective)
    Two thirds of the committee voted Republican. (noun)

3. Use a hyphen to join compound words.

    The *vice-president* spoke at the White House yesterday.
    Building *self-esteem* in children is critical to their success.
    The *well-known* author lived in Tucson.
    Mike loved living in his *great-grandmother's* home.
    The goal of each of us is to be *self-supporting.*

His *self-discipline* slipped as he responded angrily to the unfounded accusations.
The *Commander-in-Chief* of the Air Force and his family had many fond memories of the countries where they had lived.

4. Use a hyphen to join a prefix to a proper noun or proper adjective.

The company moved its headquarters to the *mid-Atlantic* states.
In *pre-Christian* times most people were polytheistic.

PARAGRAPH SYMBOL (¶)

The paragraph symbol is often used when referring to a particular paragraph in a reference book, regulation, or instruction. When used, it generally appears just before a paragraph number or identifier with no space in between.

The applicable passage is found in ¶23.22.
The reference you were looking for is in FAA Regulation 55-44, ¶43.1.5.

PARENTHESES (( ))

Parentheses are effective when used to enclose interruptions too long for a pair of commas to hold. Or, they can be used for a group of words, or a phrase, that sharply interrupts the natural word order of a sentence. Parentheses indicate the greatest degree of separation between the enclosed word group and the rest of the sentence. Punctuation within the parentheses varies according to circumstance and phrasing.

1. If the word group in parentheses occurs within a sentence, don't start it with a capital letter or end it with a period, even if the word group is a complete sentence.

My vacation (*which was way too short for me*) made my peers angry because they had to work the entire month.

2. If the word group is an exclamation or a question, put the correct punctuation mark inside the parentheses.

The softball game (*It was a runaway victory!*) gave Renee front page headline notice.

3. If you have a complete sentence that will stand alone inside the parentheses, begin it with a capital letter and end it with the appropriate punctuation mark inside the parentheses.

I went to bed late. (*Who would have known I should have started earlier?*) When the telephone rang at 6:00 A.M. to inform me that the morning meeting time had been moved to 7:30, I knew it would be a long, tiring day.

PERIODS (.)

1. Use periods at the end of sentences that neither exclaim nor ask a question, but simply make a statement.

   I think the company newspaper should come out only twice a month.

2. Use periods to follow some abbreviations and contractions.

   Dr.   Jr.   etc.   cont.   i.e.   J.D.

3. Use periods with an individual's initials.

   B.F. Goodrich   George W. Bush   F. Scott Fitzgerald

4. Use periods after numbers and letters in listings and outlines.

   Please have the following on my desk by Monday:
   1. two copies of your department needs assessment
   2. your budget outline
   3. your goals and objectives for the quarter

QUESTION MARKS (?)

Use question marks to end a direct question.

   What time does your plane arrive?

QUOTATION MARKS (" ")

Quotation marks give credit to borrowed words that you use in your own text and words that are used for special emphasis. Quotation marks visually designate directly quoted conversation, and they set off titles of short works.

1. Use quotation marks to give credit to borrowed words and words for special emphasis.

   The senior partner was a "*thorn in my side.*"
   After that presentation, I finally understand what is meant by "*seeing is believing.*"
   He thought of himself as "*King,*" but he was just an auditor.

2. Use quotations when reporting a direct statement. When commas and periods come next to quotation marks, they always come *before* the quotation marks.

> Megan Dickinson declared, "*You haven't seen the last of me!*"
> "You haven't seen the last of me!" declared Megan Dickinson.
> "*You,*" declared Megan Dickinson, "*haven't seen the last of me!*"

> Don't use quotation marks in an indirect statement:
> Megan Dickinson told our group that we hadn't seen the last of her.

3. Use quotation marks to signify titles of poems, short stories, articles, lectures, chapters of books, and television programs.

> Maya Angelou's "I Know Why The Caged Bird Sings"
> Poe's "The Tell-Tale Heart"
> Ben Dickinson's article, "1-800-Ask-Grandpa"
> Donna Ammerman's lecture on "Golf: Let Your Club Determine Your Drive Length"
> The third chapter in the history book is "The Expanding Nation."

4. A final point to remember about quotation marks is that 'single quote' marks are used when you are quoting something inside of an already existing quote.

> Dan specifically advised all employees to, "watch the stock prices go rocketing upward today, as Alan Greenspan has said there will be yet another '*significant interest rate drop*'."

SECTION SYMBOL (§)

Similar to the paragraph symbol, the section symbol denotes a particular section when referencing a document. This symbol is mostly used when quoting legal guidance or public code. Do not use a space between the section symbol and the next word or number in a sentence.

> Your client's claim is covered in §21.5 of the York County real estate code.
> The law governing corporate tax rates is in U.S. Code §10.14.5.3.

SEMICOLONS (;)

The semicolon is generally used where you could also use a period. It is best used as a kind of weak period, a separator of contrasts. Its strength comes in the joining of two sentences to form a compound sentence without tthe help of a connective word. For the semicolon to be appropriate, the sentences it joins must be closely related in meaning. If they are not, simply use a period. Semicolons can also be used instead of commas to separate items in a series when one or more items have commas of their own.

> Call me tomorrow; I will give you my answer then.
> Directions are included; they are not complete by any means.
> Seattle is a growing city; it offers a competitive business climate and a stable work-force.
> Diana's success was no surprise; she had been working hard for twenty years.
> One would think a new store would bring in more business; however, that retail market has been glutted.
> Lori speaks Portuguese, the official language of Brazil; Farsi, a Middle Eastern tongue; and Sanskrit.

Unlike commas, exclamation marks, and periods—which fall *inside* the quotation marks, colons and semicolons fall *outside* quotation marks.

> The CEO spoke of his "penchant for hiring relatives"; he might as well have just called it nepotism.

UNDERLINING AND *ITALICS*

To understand the use of underlining and italics, think about the concept that the value of a two-sided coin is the same no matter which side is showing. Underlining represents italics when you are handwriting papers, but the italic font is used when you are composing on your computer for printed material. This special font indicates titles of books, magazines, newspapers, plays, films, software, and websites. Names of specific ships or aircraft, foreign words, and words used in a special way also require italics:

> Books: *The Art of Happiness*
> Magazines: *Newsweek*
> Newspapers: *The Los Angeles Times*
> Plays: *Cats*
> Films: *Gone With the Wind*
> Software: *Power Point Presentations*
> Ships: Renaissance Cruise Lines, the *R5*
> Foreign words: *Veni, vidi, vici*
> Words used in a special way: It is difficult to say *she sells seashells by the seashore.*

VIRGULE (/)

The virgule, or forward slash, has many grammatical functions. One use is to represent the word *per* when describing units of measurement.

65 miles per hour	65 miles/hour
$8.25 per hour	$8.25/hour
Price per Earnings	Price/Earnings

Dates are abbreviated using virgules.

April 8, 1994	4/8/94
May 1966	23/5/66
December 25th	12/25

If a series of dates or numbers is used in a sentence, they can be abbreviated rather than repeating the entire expression.

1991, 1992, 1993, 1994	1991/92/93/94
extensions 4101, 4102, 4103	extensions 4101/02/03

Virgules are also used in expressing mathematical relationships or ratios.

PE ratio = Price per Earnings	PE = Price/Earnings
Velocity (V) = distance (D) over time (T)	V = D/T

Another use of virgules is in sentences using the words *and* or *or*.

He could accept the assignment *and/or* the raise.
The union demanded fewer hours *and/or* better health benefits.

► CAPITALIZATION

Rules for capitalization are straightforward, for the most part. The dictionary is a great source whenever you are in doubt, especially if you are working on a special purpose document. However, nine basic rules should be all the guide you need.

1. Capitalize the first word of every sentence and the first word of a quoted sentence.

   The CEO will organize the fundraiser for her division.
   She said, "*That* woman is a dynamo when it comes to raising money."

*If the quotation is interrupted, do not capitalize the first word of the second part of the sentence:

"That woman," she said, "is a dynamo when it comes to raising money."
Da Vinci's painting hangs in the Louvre.

*In this case, the *da* in da Vinci is capitalized because it is at the beginning of the sentence.

2. Capitalize the days of the week and months of the year, but not the seasons.

Thursday	tomorrow
February	winter
Easter	spring
Labor Day	fall

3. Capitalize proper nouns. Remember, proper nouns name a specific person, place, or thing. This includes all registered trademarks.

New England Patriots	George W. Bush
Honolulu, Hawaii	Golden Gate Bridge
Bayer Aspirin	Swedish pancakes
Revolutionary War	Battle of the Bulge
Kleenex	Xerox

4. Capitalize a person's title when it precedes the proper name. Do not capitalize when the title is acting as a description following the name. Capitalize the title when it stands in place of the name.

Chairperson Koblitz will conduct the meeting.

Sincerely,
Mary Koblitz, Chairperson

Mary Koblitz, chairperson of the committee, will conduct the meeting.
Lori met General Worden in Paris for the meeting.
They invited Senator Craig to attend, as well. The media group shouted, "Welcome home, General*!"

General takes the place of the whole name, General Worden.

5. Capitalize the titles of high-ranking government officials when used with or without their names.

> The children invited President Bush to speak at their school.
> The President spoke eloquently and the crowd responded enthusiastically.
> The Senators gave the visiting head of state a standing ovation.

6. Capitalize all key word words in the titles of books, poems, paintings, etc. (note that "of" is not capitalized in the first title).

> *The Seven Habits of Highly Successful People*
> *The Star Spangled Banner*
> *The Mona Lisa*

7. Capitalize the regions of the country, but do not capitalize directions or points on a compass.

> Donna says that the West is where the company could really make strides.
> The chief accountant traveled east to visit other sites.
> She finally decided the East would be a better location.

8. Capitalize the first word of a salutation and the first word of the closing of a letter.

> Dear Mrs. Benoit: Very truly yours,

9. Capitalize acronyms, abbreviations for governmental agencies, other organizations, and corporations.

> Intel Corporation
> NSA
> OAS
> AR (Action Requested)

▶ SPELLING

You probably thought—or at least you hoped—that you had left spelling issues back in the third grade. But, alas! They remain. Mastering them is more important than ever in your business correspondence. It comes down to this: If you can spell correctly, you appear smarter. This is a good thing.

Luckily, in our age of technology, we enjoy spellcheckers as part of most word processing systems. Spellcheckers catch all of the obvious typos and blatant misspellings, thus allowing for very effective and easy editing. However, even though your spellchecker is an extremely useful tool, it does not solve every spelling issue. You still need to *proofread* your own letters. And, you still need to adhere to some basic spelling ground rules in order to be consistently polished and correct in your letters.

You can proofread by skimming your index finger along each word, whispering aloud to yourself as you go. Or, you can go over your text backwards, which enables you to focus on each individual word, rather than on the content of the letter. Be sure to check and double-check the spelling of all proper names and company names as well, as your spellchecker won't do that for you either.

The bottom line is that you can make spelling an easier issue for yourself by following some basic spelling ground rules. The following spelling section offers some helpful spelling basics, and a long list of commonly misspelled words. Memorize the words that plague you most often, and look up the rest in a dictionary as needed. Don't forget to look over your whole document one last time for any "commonly confused" words. Always keep in mind that your letter is most effective when you use the right words in the right context.

SPELLING GROUND RULES

- Use your spellchecker if you have one.
- Use a dictionary—keep a small pocket dictionary on hand to quickly look up tricky words.
- Say the "problem" word aloud to yourself. Does it look right? Does it sound right?
- Read the entire text backwards to focus on each individual word.
- Learn some spelling tips and tricks (see list below).
- Memorize the words you most often misspell (see list ahead).
- Proofread your document—again, your spellchecker will not catch all proper names or typographical errors (typos).
- See the list of commonly confused words in Chapter 1, page 37.

Basic Spelling Tips:

1. If spelling is a challenge for you, consider using shorter words. Long, multi-syllabic, hard-to-spell words often come across as pretentious. Keep your language simple and straightforward and focus on the message, not the vocabulary.

    Remember to put *i* before *e*, except after *c*.

| categories | dossier | grieve | mischievous | piece | territories |
| conceit | conceive | deceit | deceive | receipt | receive |

2. There are three rules for words ending in *ed* or *ing*. If a word ends in a <u>vowel</u>, drop the vowel and add *ed* or *ing*.

appreciate	appreciated	appreciating	confine	confined
confining	desire	desired	desiring	gravitate
gravitated	gravitating	negate	negated	negating
surprise	surprised	surprising		

3. If a word ends with a <u>consonant</u>, but the letter before the consonant is another consonant, simply add *ed* or *ing* to the end of the word.

back	backed	backing	hang	hanged	hanging
guard	guarded	guarding	mark	marked	marking
rent	rented	renting	thwart	thwarted	thwarting

4. If a word ends with a <u>consonant</u>, but the letter before the consonant is a vowel, then double the ending consonant and add *ed* or *ing*.

drop	dropped	dropping	grab	grabbed	grabbing
hop	hopped	hopping	knit	knitted	knitting
ship	shipped	shipping	trap	trapped	trapping

5. One last suggestion is to make a list of those words you continue to struggle to spell correctly. After enough references to this individualized list, you will have memorized many of these challenging words.

COMMONLY MISSPELLED WORDS

acceptable	exhibition	mnemonic	salable
accommodate	exonerate	mortgage	satellite
achievement	exorbitant	necessary	secretary
acquisition	facsimile	negotiate	sergeant
advantageous	financially	obsolescent	significant
affect	foreign	omitted	simultaneous
affidavit	fulfill	opportunity	sizable
amortize	government	optimism	skeptic
analysis	guarantee	ordinarily	skillful
answer	hindrance	paradigm	spacious
appearance	illicit	parallel	stationary/stationery
appropriate	immediately	particularly	statistics
arbitrary	independent	performance	strategy
argument	inequity	permissible	strenuous

attorney	inevitable	perseverance	subpoena
auxiliary	influential	personnel	subtlety
bankruptcy	initiative	phase	superintendent
benign	innuendo	phenomenon	surgeon
boundary	insistence	phony	surveillance
bureaucracy	interim	possession	synonymous
cancellation	interrupt	precedence	tariff
characteristic	irrelevant	prerequisite	taxing
chronological	itinerary	prerogative	technique
committee	judgment	pretense	temporary
competent	khaki	prevalent	tempt
conference	ledger	principle	theory
consensus	legitimate	procedure	thorough
deductible	liable	process	threshold
defendant	liaison	prohibition	transferable
deficit	license	propaganda	unanimous
definite	lien	protege	undoubtedly
development	lieutenant	quantity	unmanageable
dignitary	logistics	questionnaire	unwieldy
dissimilar	maneuver	recession	usually
dossier	martyr	recognize	various
effect	mediocre	reference	visible
elicit	memento	reinforce	volume
emphasize	millennium	relevant	warehouse
entrepreneur	millionaire	relieve	warrant
equivalent	minor	repetition	whether
escrow	miscellaneous	restaurant	wholly
exceed	misspell	rhetorical	yield

SUMMARY

So, there you have it. You now know all of the grammar basics that you will ever need to know in order to write successfully in business. Remember that carefully crafted, correct sentences, proper structure, and good spelling are the most effective means of communication there is. All you need to do is come up with quality content.

Never forget that your writing represents you, and that you can become the "Michelangelo" in all that you write with a little forethought and some solid skills. Hopefully, this book has opened your mind to the "artist" that lives within you.

Time to get down to business!

Master the Basics... Fast!

If you need to improve your basic skills to move ahead either at work or in the classroom, then our LearningExpress books are designed to help anyone master the skills essential for success. It features 20 easy lessons to help build confidence and skill fast. This series includes real world examples—**WHAT YOU REALLY NEED TO SUCCEED.**

All of these books:

- Give quick and easy instruction
- Provides compelling, interactive exercises
- Share practical tips and valuable advise that can be put to use immediately
- Includes extensive lists of resources for continued learning

---

Write Better Essays
208 pages • 8 1/2 x 11 • paper
$13.95 • ISBN 1-57685-309-8

The Secrets of Taking Any Test, 2e
208 pages • 7 x 10 • paper
$14.95 • ISBN 1-57685-307-1

Read Better, Read More, 2e
208 pages • 7 x 10 • paper
$14.95 • ISBN 1-57685-336-5

Math Essentials, 2e
208 pages • 7 x 10 • paper
$14.95 • ISBN 1-57685-305-5

How To Study, 2e
208 pages • 7 x 10 • paper
$14.95 • ISBN 1-57685-308-X

Grammar Essentials, 2e
208 pages • 7 x 10 • paper
$14.95 • ISBN 1-57685-306-3

Improve Your Writing For Work, 2e
208 pages • 7 x 10 • paper
$14.95 • ISBN 1-57685-337-3

To Order: Call 1-888-551-5627

Also available at your local bookstore. Prices subject to change without notice.

LearningExpress • 900 Broadway, Suite 604 • New York, New York 10003

LEARNINGEXPRESS®

LearnATest.com™